BASIC ELECTRONICS

BASIC ELECTRONICS

S.Y. KULKARNI

Principal

NMAM Institute of Technology

Nitte

Karnataka

B. SOMANATHAN NAIR

Principal, SHM College of Engineering and Technology, Kadakkal

Thiruvananthapuram

and

Visiting Professor, Department of Optoelectronics

University of Kerala, Thiruvananthapuram

TECHSAR™

NEW DELHI

Published and Printed by

Techsar Pvt. Ltd.
4435-36/7, Ansari Road, Daryaganj
New Delhi-110 002 (India)
E-mail: tech_info@techsarworld.com
Website: www.techsarworld.com

ISBN: 978-93-80578-81-1

Reprint 2024

Dedicated to my dear friend

Adv. V.S. Hareendranath

by

B. Somanathan Nair

PREFACE

This is intended as a textbook to cater the needs of first-year students of all the branches of engineering.

The First Chapter describes semiconductor diodes and their applications in engineering. The discussions start with an introduction to the *PN* junction diode and its forward and reverse characteristics. This is followed by a discussion on diode approximations and the DC load line. We next discuss the temperature dependence of the characteristics, followed by the AC equivalent circuit of the diode. After this, we explain rectifiers and Zener diodes.

In the Second Chapter, we explain the theory and applications of various configurations of bipolar junction transistors as amplifiers. DC and AC load lines are also explained in this unit.

In Chapter 3, we discuss various types of biasing of transistors. This is followed by stability-factor analysis.

Chapter 4 deals with various electronic devices such as SCR, UJT and JFET. The structures and characteristics of these devices are also explained.

Amplifiers and oscillators are discussed in Chapter 5. Discussions begin with an explanation on the units of bel and decibel. This is followed by discussion on the theory of single-stage and two-stage RC-coupled amplifiers. This unit finds some brief discussion on negative voltage series feedback and its effects. Next we explain the general theory of oscillations and various types of oscillators.

In Chapter 6, we have a brief encounter with operational amplifiers. This is followed by a description of cathode-ray oscilloscope.

In Chapter 7, we make a thorough discussion on various types modulations and radio transmitters and receivers. In this same unit, we also discuss various types of binary systems.

In Chapter 8, we go through the basics of Boolean algebra, logic gates and binary adders.

In all the units, we have added a number of solved examples to illustrate the theory explained. At the end of each chapter, we have also given a large number of one-word/multiple choice questions, review questions and numerical problems. The answers to one-word questions and numerical problems are given at the end of the text. The student is advised first to work out the problems and then refer to the answers.

The book is written to explain the complex theories in a very simple manner so that any reader with basic knowledge on physics can easily grasp them. It is hoped that the book will be very helpful to all the students.

S.Y. Kulkarni

B. Somanathan Nair

CONTENTS

5. Amplifiers and Oscillators 91

Semiconductor Diodes and Applications

CHAPTER 1

1.1 PN JUNCTION DIODE

Figure 1.1 shows the structure an ordinary *PN* junction diode. As shown in Fig. 1.1, the diode usually consists of a $P^{++}NN^{++}$ structure. The heavy dopings (P^{++} and N^{++}) are required to avoid the formation of Schottky-barrier junctions. It may be noted that Schottky-barrier junctions are metal-semiconductor junctions, which are formed naturally when aluminium is deposited on the semiconductor to take out external connections. It is seen that these junctions act as rectifiers. However, such junctions can be avoided if we introduce heavy doping by diffusion (say, N^{++}) in between a lightly doped semiconductor (say, N) and the metal contact. This is because, N and N^{++} regions are both semiconductors of the same substance, and hence they can form a non-rectifying junction. Similarly, N^{++} semiconductor is almost like a metal in all aspects; therefore, it can form a non-rectifying contact with the cathode metal connection. Therefore, with the NN^{++} - metal structure, we get non-rectifying junctions at places where junctions of this kind have to be avoided. In the same fashion, we find that the P^{++} region and the anode metal contact form non-rectifying junctions. Thus we find that, in the conventional *PN* junction diode, junction exists only between the N and P^{++} regions. After the external leads are attached, the whole unit is encapsulated in a plastic or metal opaque covering.

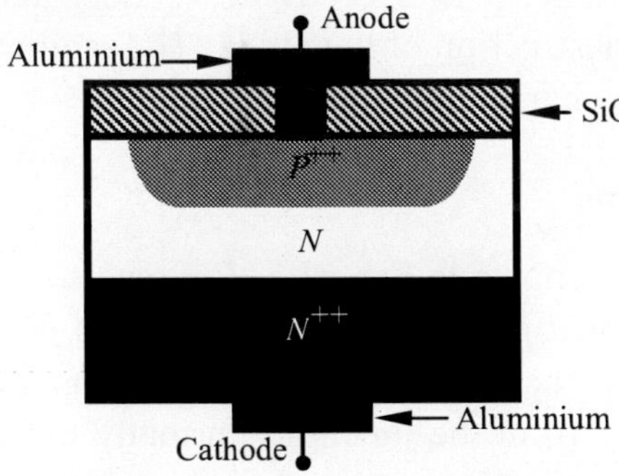

Fig. 1.1 Conventional *PN* junction diode.

1.2 CHARACTERISTICS OF PN JUNCTION DIODES

1.2.1 Diode under Forward-Bias

Consider a silicon *PN* junction diode (1N 4001), whose anode is connected to the positive terminal and cathode is connected to the negative terminal of a battery (V_{AA}), as shown in Figs. 1.2. This operation is known as *forward biasing* of a diode. Figure 1.2(a) shows a block model and Fig. 1.2(b) shows the same circuit drawn with the diode represented by its symbol.

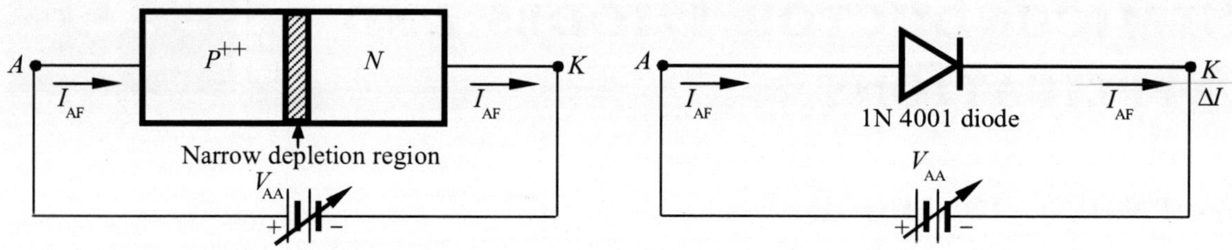

(a) Diode (block model) (b) Circuit diagram using symbol

Fig. 1.2 Forward biasing of *PN* junction diode.

Let V_{AA} be varied from 0 to 1 V in steps of 0.1 V each. It is a well known fact that across any *PN* junction, there exists a built-in potential (called the *junction potential* V_J) due to the immobile acceptor and donor ions present near the junction. When the externally applied anode-cathode voltage V_{AK} (= V_{AA}) exceeds V_J, majority-carrier holes from the *P* region and majority-carrier electrons from the *N* region cross the junction and forward anode current flow takes place through the diode. The value of this external potential, which excites current flow through a given *PN* junction diode, is called the *cut-in voltage* V_C of that diode. V_C is approximately 0.5 V for silicon diodes.

As stated above, once $V_{AK} \geq V_C$, forward anode current I_{AF} begins to flow through the diode. It is observed from the forward *VI* characteristics shown in Fig. 1.4 that I_{AF} increases in an exponential manner. When V_{AK} is about 0.8 V, very high anode current flows through the diode and we say that the diode is in a state of *saturation*. When V_{AK} exceeds 0.8 V, abnormally high anode current flows, which eventually leads to the total destruction of the diode. The region between 0.5 V and 0.8 V is called the *active region* of the diode characteristic.

1.2.2 Diode under Reverse Bias

Let us now reverse bias the diode, as shown in Figs. 1.3. For reverse biasing the diode, we connect *P* of the diode to the negative terminal of the battery and *N* to its positive terminal. In this case, we find that holes are attracted by the battery-positive and electrons by the battery-negative. If the holes and electrons are pulled away from the junction, normally there should be no current flow through the diode. However, we observe that under reverse bias there exists a small reverse current I_R, which is made up of two components.

a) Component of reverse current I_R due to minority carriers

We know that electrons are minority carriers (much smaller than majority the carriers, i.e., holes) in the *P* region. Similarly, holes are minority carriers in the *N* region. For example, let the number of holes be 10^{17} per cm^3 in a *P* type silicon sample, then the number of electrons in that sample can be found to be only 10^3 per cm^3. With the applied reverse bias, these can cross the junction to produce a small reverse current. This component of the reverse current is known as the *reverse-saturation current* I_0. It is a constant for a given temperature due to the fact that the number of electron-hole pairs (EHPs) produced for a given temperature is a constant, and I_o is produced due to the flow of these EHPs. Since I_o is constant, it is given the name *saturation current.* The number of minority carriers is very small in Si at room temperature. I_o is usually in the pico- or nano-ampere range. This component is shown in Fig. 1.5.

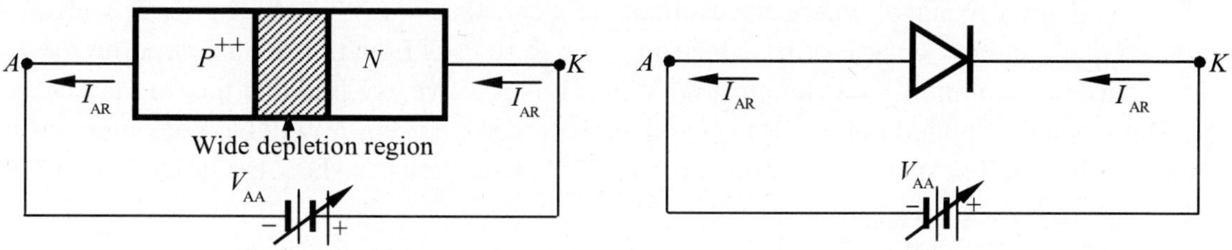

(a) Block model (b) Circuit diagram using symbol

Fig. 1.3 Diode under reverse bias.

b) Component of reverse current due to surface leakage

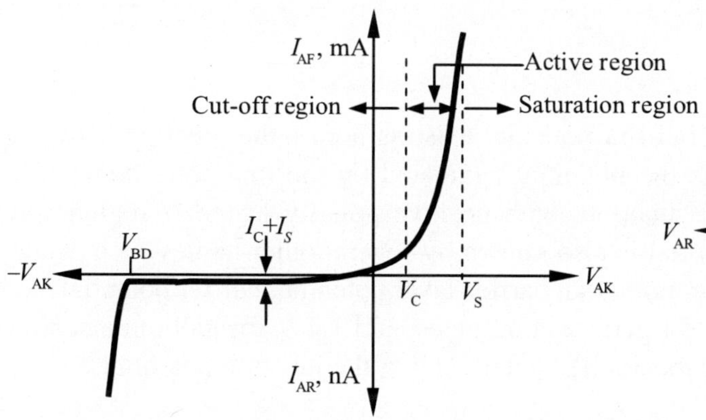

Fig. 1.4 *V-I* characteristics of a typical *PN* diode.

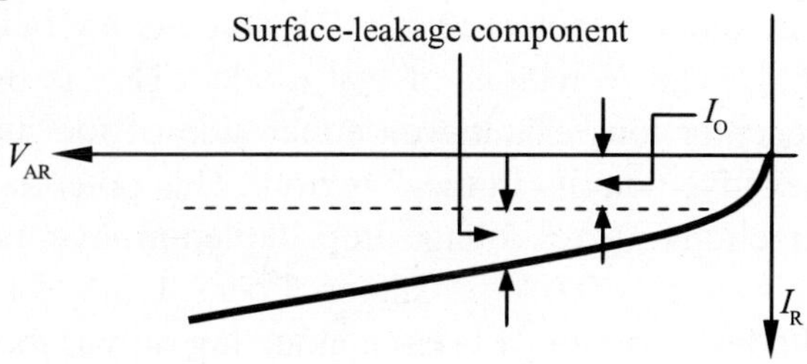

Fig. 1.5 Expanded view of reverse current.

We have already seen that there exists unsatisfied bonds near the surfaces of semiconductors. These bonds will give free electrons for conduction on application of an electric voltage. This is a *drift current* and obeys the Ohm's law. So, we notice that *this component of reverse current increases with increase in the applied reverse voltage.* Usually, for a well-prepared semiconductor surface, this leakage current will also be in the pico- or nano-ampere range. This component, superimposed over I_o, is also shown in Fig. 1.5.

Even though the reverse current is in the pico- or nano-ampere range, we find that in certain situations this can suddenly become quite large. When the reverse current becomes abruptly high, keeping the reverse voltage at a *constant* value, we call the situation a *reverse breakdown*. Breakdown occurs when the reverse voltage exceeds certain finite values, which in turn depends on the structure of the diode. Reverse breakdown is explained more in Section 1.8.

From analysis, we can express the diode current I by the equation:

$$I = I_R \left(\epsilon^{V/\eta V_T} - 1 \right) \tag{1.1}$$

where I_R = diode reverse current, V = applied voltage, and $\eta = 2$ for silicon and 1 for germanium. $V_T = kT/e$ is thermal voltage, where k = Boltzmann's constant = 1.38×10^{-23} J/K, T = absolute temperature in K, and e = charge of the electron = 1.6×10^{19} C. Equation 1.1 is based on the fact that diode current is a minority-carrier current. When V is positive, we find that the forward current is high. But when the applied voltage V is negative, the reverse current I_R will be quite small, as the exponent will be small in such cases. The combined VI characteristics show the breakdown region also.

1.3 PARAMETERS OF A PN JUNCTION DIODE

The parameters associated with a *PN* junction diode are the junction potential, the static forward resistance, the dynamic forward resistance, the peak-inverse voltage PIV, the reverse current and the maximum forward current.

1.3.1 Junction Potential V_J

The junction potential of a *PN* diode is the built-in potential existing across the junction between the *P* and *N* regions of that diode. This is the potential generated by the immobile donor and acceptor ions existing on either side of the junction, with positive polarity in the *N* region and negative polarity in the *P* region. This potential is also known by several other names such as the junction forward voltage drop, depletion-layer potential, barrier-layer potential, barrier potential etc. It is roughly 0.6 V for silicon diodes, 0.3 V for germanium diodes, and 1.4 V for gallium arsenide diodes. A diode will start conducting only if the applied potential is ≥ the junction potential.

1.3.2 Static Forward Resistance R_F

Static forward resistance is the resistance of the diode for a given value of anode-cathode forward voltage V_{AF} and the resulting anode-cathode forward current I_{AF}. From Fig. 1.6(a), we obtain the static forward resistance from the computation

$$R_F = V_{AF}/I_{AF} \tag{1.2}$$

1.3.3 Dynamic Forward Resistance r_f

Dynamic forward resistance is the resistance of the diode for a given change in the anode-cathode forward voltage (ΔV_{AF}) and the resulting change in the anode-cathode forward current (ΔI_{AF}). From Fig. 1.6(b), we obtain the dynamic forward resistance from the computation

$$r_f = \Delta V_{AK}/\Delta I_{AF} \tag{1.3}$$

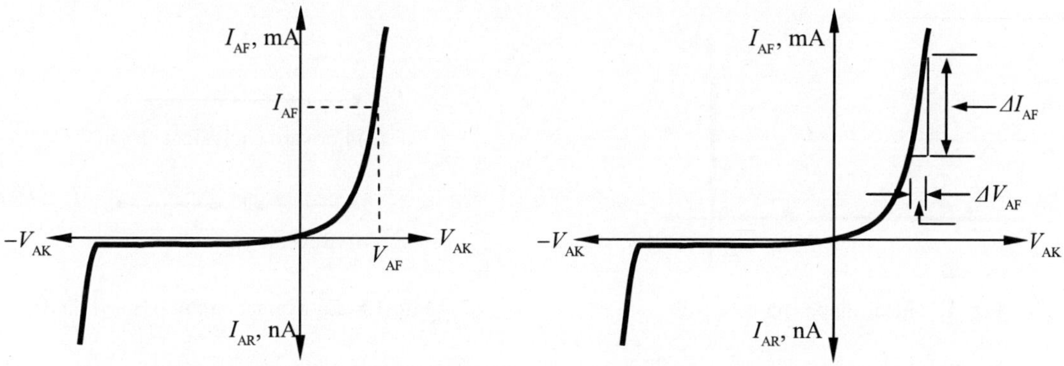

(a) Computation of static forward resistance R_f (b) Computation of dynamic forward resistance r_f

Fig. 1.6 Computation of static and dynamic forward resistances in *PN* diodes.

1.3.4 Peak-Inverse (reverse breakdown) Voltage V_{BR}

Peak-inverse voltage (PIV, as this commonly called) is the maximum value of the voltage that can be applied across the diode in the reverse direction *without breaking it down.* PIV is one of the most important parameters of the diode, when it is used as a rectifier. It is also one of the parameters that determine the cost of a diode.

1.3.5 Maximum Forward Current $I_{AF(max)}$

This is the maximum value of the forward current that is permitted through a diode before breakdown.

1.3.6 Average Forward Current I_{AF}

This is the optimum safe value of the forward current that is permitted through the diode. Any diode circuit is designed so that the current flowing through it will be normally I_{AF}.

1.3.7 Reverse Saturation Current I_{AR}

As stated earlier, this is the value of the current that will be flowing through a diode under the reverse-bias condition and will be usually in the nano- or pico-ampere range for silicon diodes.

1.4 DIODE APPROXIMATIONS (IDEAL AND PRACTICAL DIODES)

Figure 1.7 shows the characteristic of an ideal diode. Ideal diode is an ideal switch. It will conduct with zero resistance in the forward direction and will be open in the reverse direction. Thus the characteristic coincides with the y-axis in the forward direction (showing zero forward resistance) and the x-axis in the reverse direction (showing infinite reverse resistance). Figure 1.8(a) shows the equivalent circuit of the ideal diode in the forward direction and Fig. 1.8(b) shows its reverse equivalent circuit.

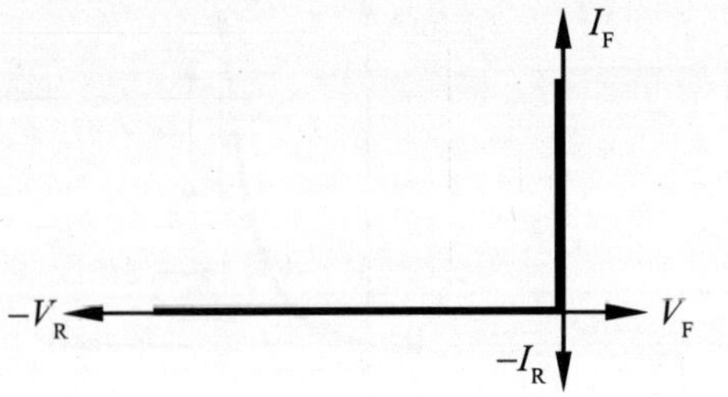

Fig. 1.7 Ideal-diode characteristic.

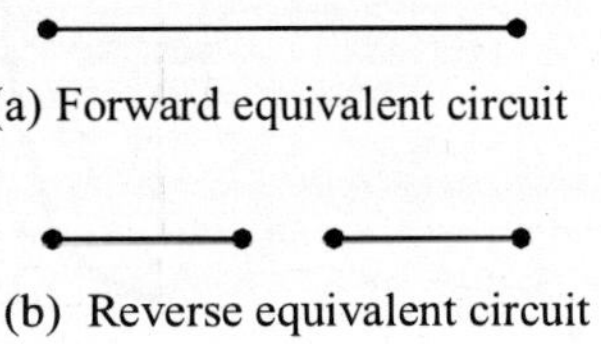

Fig. 1.8 Equivalent circuit of ideal diode.

Figure 1.9 shows the characteristics of a practical diode. A practical diode can be seen to exhibit a small amount forward resistance r_f (about 10 to 100 ohms) and a small cut-in voltage V_D (of about 0.6 V for silicon diodes), as shown in Fig. 1.9. The DC equivalent circuits of a practical diode in the forward and reverse directions are shown in Fig. 1.10(a) and (b), respectively.

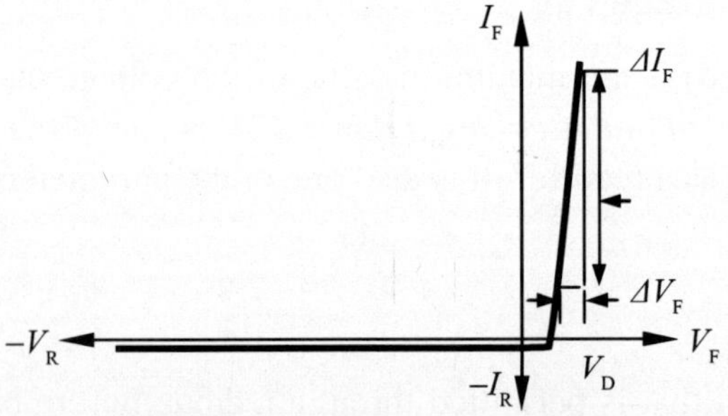

Fig. 1.9 Practical-diode characteristic.

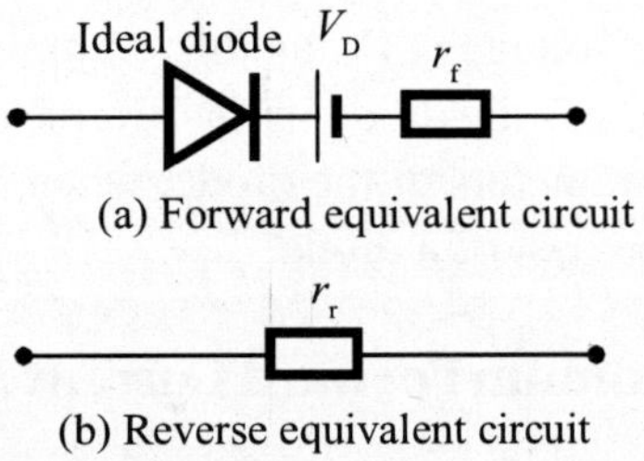

Fig. 1.10 Equivalent circuit of practical diode.

1.5 DC LOAD-LINE ANALYSIS

Consider a *PN* diode *D* under forward bias, as shown in Fig. 1.11. Figure 1.12 shows its forward characteristic. Let V_{AA} be the maximum applied bias and let R_L be the load resistance. From Fig. 1.12, we find that at V_{AA}, $I_F = 0$. Similarly, at $V_{AA} = 0$, $I_F = I_{F\,max}$ $(= V_{AA}/R_L)$. The line joining V_{AA} and V_{AA}/R_L is called the **DC load line**. The slope of this line is $1/R_L$.

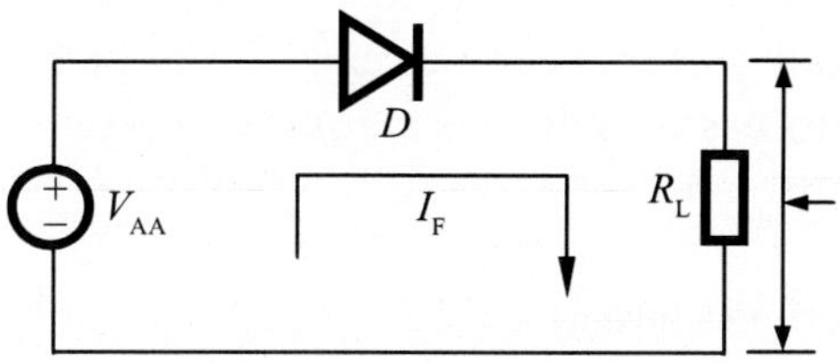

Fig. 1.11 Diode connected to a load.

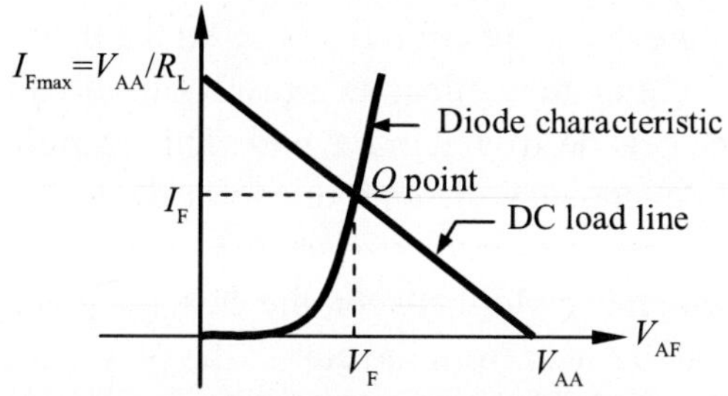

Fig. 1.12 Plotting of DC load line.

From Fig. 1.12, we notice that the load line cuts the *VI* characteristics at a point called the *Q* or *operating point*. The voltage corresponding to this point V_F is the voltage drop across the diode and the current corresponding to this point I_F is the current flowing through the diode for the given value of V_{AA} and R_L. We may now express the current-voltage relation in the circuit by the expression

$$V_{AA} = I_F R_L + V_F \tag{1.4}$$

Example 1. Compute the value of V_F and I_F given that a diode circuit consists of $V_{AA} = 4.0$ V, $R_L = 0.8$ kΩ, and a characteristic as shown in Fig. 1.13(a). Compute the values of V_F and I_F.

Fig. 1.13(a) *V-I* characteristics of a typical *PN* diode.

Fig. 1.13(b) DC load line drawn on characteristic.

Solution

We first draw the DC load line on the characteristic. For drawing the load line, we join the points $V_{AA} = 4.0$ V and $I_{F\,max}$ (= 4 V/0.8 kΩ) = 5 mA, as shown in Fig. 1.13(b). We find that the DC load line cuts the characteristic at the *Q* point defined by the co-ordinates $V_F = 0.7$ V and $I_F = 4$ mA. This suggests that with a load resistance of $R_L = 0.8$ kΩ and an applied bias $V_{AA} = 4.0$ V, the diode will have a normal current of 4 mA flowing through it producing a voltage drop of 0.7 V across it.

1.6 TEMPERATURE EFFECTS

All semiconductor devices are temperature dependent. When the ambient temperature rises, diode current increases due to the increase in the number of electron-hole pairs. This in turn increases the temperature of the junction resulting in the production of more EHPs and hence more diode current. This is a cumulative effect as a result of which the diode current becomes excessively high so that the diode gets destroyed in the end. This cumulative process is known as *thermal runaway.*

The dependency of semiconductor devices on temperature are expressed in the form of a curve known as the *temperature de-rating curve.* Figure 1.14 shows a typical de-rating curve of a *PN* diode, which is a plot between the case (or junction) temperature of a diode.

As can be seen from the curve, the power dissipated at the junction of the diode is a constant of value P_D and within safe limits upto a temperature of $T = T_A$°C. Usually, T is taken as 25°C. The difference in the junction temperature (T_J) corresponding to power dissipated P_D and the ambient temperature (T_A) can be seen to be proportional P_D. This relation can be expressed in the mathematical form

$$(T_J - T_A) \propto P_D \tag{1.5}$$

This relation is similar to the Ohms law and hence is sometimes called as the *Thermal Ohms law*. The relation can be completed by expressing the relation in the form

$$\theta_J = \Delta T_j / \Delta P_D \ \ \text{T/°C} \tag{1.6}$$

where θ_J is called the *thermal resistance* of the junction. The inverse of thermal resistance is called the de-rating factor *D*, which may be expressed as

$$D = \Delta P_D / \Delta T_j \ \ \text{W/°C} \tag{1.7}$$

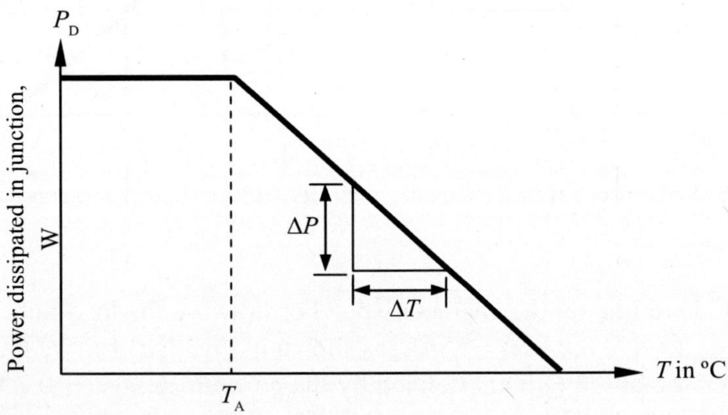

Fig. 1.14 Temperature de-rating curve of a *PN* diode.

It can be seen from Fig. 1.14, as the junction temperature increases, the power dissipation capacity of the diode decreases. It is thus advised that the ambient temperature must be less than the safe value suggested (i.e., around 25°C). If the temperature of the junction increases above this, we may use heat sinks to dissipate the power away to the surroundings.

1.7 AC EQUIVALENT CIRCUITS

Figures 1.15(a) and (b) show the equivalent circuit of a *PN* diode operating under AC signalling condition in the forward-biased mode. In this mode, a practical *PN* diode is represented by the ideal diode in series with its small dynamic forward resistance r_f (of a few tens of ohms). In parallel to this combination, we find that there exists a parasitic capacitance across the diode. This capacitance is actually made up of two capacitances, viz., the junction capacitance and the diffusion capacitance C_D.

The junction capacitance C_J is made up of the immobile donor and acceptor ions existing on either side of the junction and is usually a few picofarads in value. The diffusion capacitance C_D is made up of the mobile charge carriers (electrons and holes) that move towards the junction and accumulate their under the forward-bias mode creating the effect of a capacitance. Since these charge carriers are large in number, C_D will be about a few hundreds of picofarads (it may be noted that C_D exists only during the forward-bias condition). Since $C_J \ll C_D$, it is usually neglected from calculations.

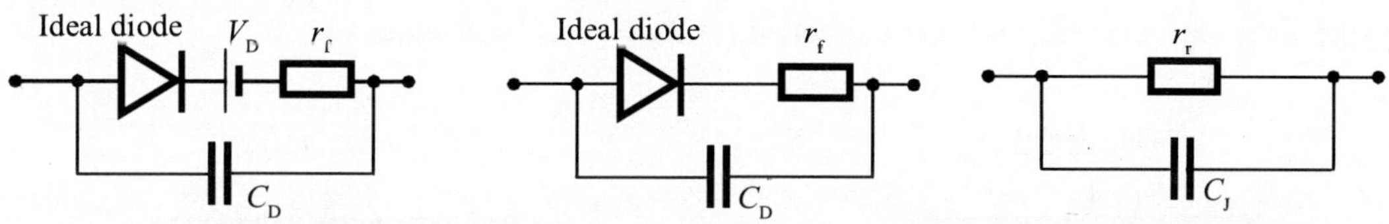

(a) Forward equivalent circuit (b) Simplified forward equivalent circuit (c) Reverse equivalent circuit

Fig. 1.15 AC Equivalent circuit of a practical diode.

Figure 1.15(c) shows the equivalent circuit of a *PN* diode operating under AC signalling conditions in the reverse-biased mode. In this mode, the equivalent circuit is just the large reverse resistance r_r (of a few thousands of ohms) in parallel to the junction capacitance C_J. It may be noted that $C_D = 0$ during the reverse-biased mode, as the mobile charge carriers move away from the junction under reverse bias.

1.8 BREAKDOWN DIODES

When the reverse voltage reaches some specified value (V_{BD}), we find that the reverse current rises abruptly, keeping the reverse voltage constant at that value, as shown in Fig. 1.16. This situation was called reverse breakdown. It is found that *reverse breakdown* can occur in semiconductor diodes by two entirely different mechanisms: the avalanche breakdown and the Zener breakdown.

1.8.1 Avalanche Breakdown or Impact Ionisation

This occurs in the conventional *PN* or P^+N diodes. In these diodes, the junction depletion-layer width is quite large, as shown in Fig. 1.16(a). When the reverse voltage V_R is high (> 8 V), the charge carriers (EHPs) acquire sufficient kinetic energy and strike the parent atoms to break the covalent bonds. The EHPs so generated immediately attain enough kinetic energy and strike other atoms to produce additional EHPs. In this way, within a short period of time, a very large number of EHPs are produced and the reverse current suddenly increases from its low value to very high value. A peculiarity of this is that the reverse voltage across the diode, V_R, remains a constant at this break-down value. As this resembles the *avalanche* occurring in ice hills, this breakdown is called the *avalanche breakdown.* Since it is produced due to the collision of electrons and atoms, it is also known as *impact ionisation.*

1.8.2 Zener Breakdown or Field Ionisation

This occurs in diodes, which are formed due to heavy doping. When a diode is constructed with heavy impurity doping (doping level > 10^{19} impurity atoms/cm^3) to form a P^{++}-N^{++} structure, the junction barrier thickness becomes very small, as shown in Fig. 1.6(b). In this case, the electric field created across the junction will be extremely high even if the applied voltage is quite small. This can be seen from the expression for the electric field $\epsilon = V/d$ volt/meter.

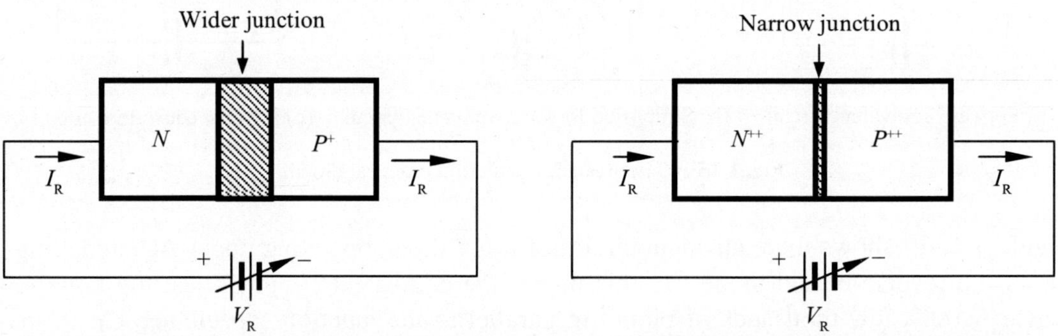

(a) Wide junction - Avalanche action (b) Narrow junction - Zener action

Fig. 1.16 Avalanche and Zener breakdowns in *PN* junctions.

With $V = 1$ V, and $d = 10^{-4}$ cm, we find using the above equation that $\epsilon = 10^4$ V/cm. It is a well known fact that, in many cases, if $\epsilon \geq 10^4$ V/cm, then dielectric breakdown will occur because of the *inability of the dielectric to withstand high electric fields.* This results in the generation of large number free charge carriers (electrons and holes) which in turn produce an extremely large current. Therefore, we find that in heavily doped semiconductor diodes, due to field ionisation, large reverse currents are produced at low reverse voltages. This mechanism is known as *Zener breakdown*, named after its inventor Carl Melvin Zener, an American Scientist. Zener breakdown is seen to occur in heavily doped PN junction diodes with applied reverse voltages of less than 5 V. A comparison between the avalanche and Zener breakdowns is shown in Table 1.1.

Table 1.1 Comparison between ABD and ZBD

Sl.No.	Avalanche Breakdown (ABD)	Zener Breakdown (ABD)
1.	Junction thickness is large (> a few μ)	Junction thickness is small (< a few μ)
2.	Conventional P^+N or P^-N Junction	Heavily doped $P^{++}N^{++}$ diode
3.	Impact Ionisation	Field Ionisation
4.	Occurs for reverse voltages > 6 V	Occurs for reverse voltages < 5 V

1.8.3 Characteristic of Breakdown Diodes

The characteristic of the breakdown diode is shown in Fig. 1.17. This is an extension of the characteristic of the ordinary diode, shown in Fig. 1.6. The curve in the reverse direction is of more importance in for the breakdown diodes than that in the forward direction. It may be noted that the curve in the forward direction is similar to that of the ordinary diode.

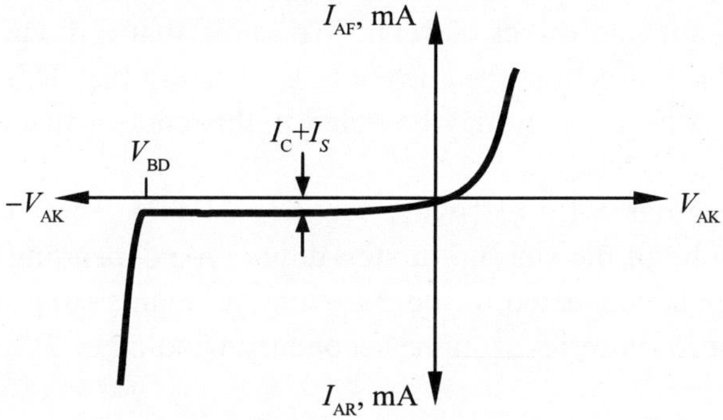

Fig. 1.17 Characteristics of a typical avalanche diode.

1.8.4 Construction of Zener and Avalanche Diodes

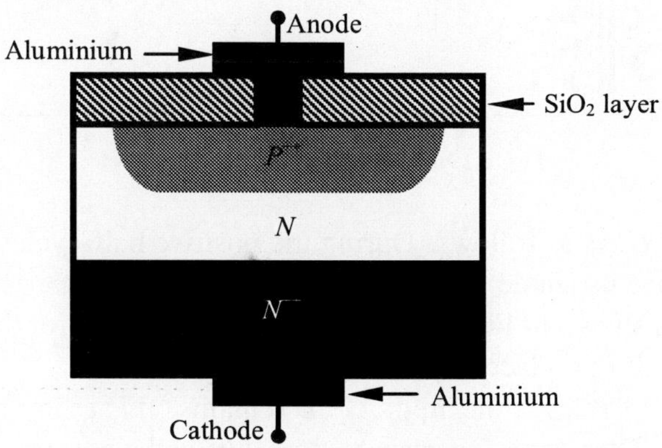

Fig. 1.18 Avalanche diode.

The construction of the avalanche diode is similar to that of the conventional $P^{++}N$ diode, designed with special care so that breakdown occurs at desired voltages. Such a structure is shown in Fig. 1.18.

For the Zener diode, the doping levels are to be adjusted appropriately. Another aspect to be taken into account to design the device is that, when the diode is to be operated as a breakdown diode, it should be able to operate in the breakdown region infinite number of times without getting destroyed. For this, the diode structure must be modified to withstand large currents that will be produced during the breakdown. In power Zener diodes, we add heat sinks to the anode to dissipate the heat generated in the junction.

1.9 APPLICATIONS OF CONVENTIONAL PN DIODES

1.9.1 Half-Wave Rectifier

One of the important applications of ordinary *PN* diodes is in rectifiers, where they are used for converting alternating current into direct current. We know that a diode conducts heavily when forward biased. In the reverse direction, the current flow is negligible. This property is made use of in rectifiers, or AC-to-DC converters. It may be noted in this context that a DC-to-AC converter is called an *inverter*.

Figure 1.19 shows the circuit of a half-wave rectifier (HWR). The circuit consists of a power transformer T, which may be of the step-up or step-down type depending on the application. The primary of the transformer is connected to the 230-volt AC mains supply. Finally, we connect a diode D and a load resistor R_L in series with the secondary winding of T, as shown in Fig. 1.19.

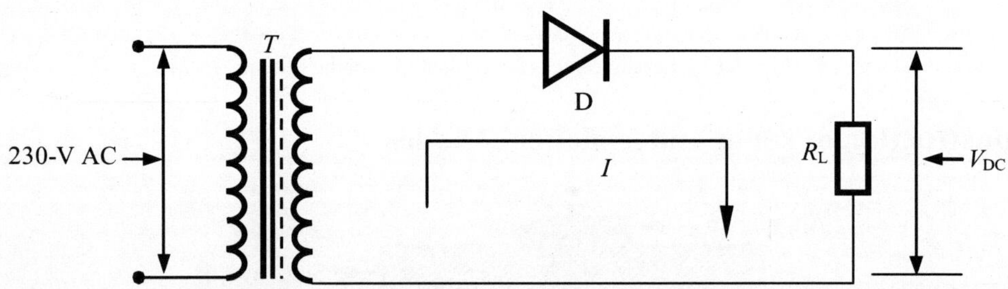

Fig. 1.19 Half-wave rectifier circuit.

The working of the HWR is as follows. During the positive half-cycles (phc's), D will conduct in the direction shown, since its anode is more positive than its cathode. This permits a current I to flow through the circuit, as shown in Fig. 1.19. The equivalent circuit for this condition is shown in Fig. 1.20(a), where the diode is replaced by an ON switch.

During the negative half-cycles of the input AC, D remains OFF, as its anode is negatively biased with respect to its cathode. Therefore, I_A ceases to flow during the negative half-cycles (nhc's). The equivalent circuit for this condition is shown in Fig. 1.20(b), where the diode is replaced by an open switch.

Thus, we observe that *there will be current flow only during the positive half-cycles* in a HWR. This is equivalent to saying that the alternating current has been converted into a direct current by the diode. It should, however, be noted that this is a variable or pulsating DC, and not a steady (constant) one. To make this variable DC into a steady one, we have to use filters.

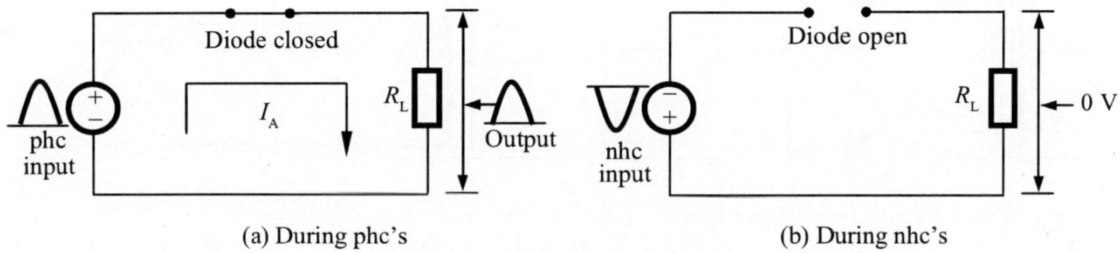

Fig. 1.20 Working of HWR.

The waveforms associated with the half-rectifier are shown in Figs. 1.21. From Fig. 1.21(a), we notice that the output is similar to the input during the positive half-cycles. However, from Fig. 1.21(b), we find that there is no current flow during the negative half-cycles. Thus we find that the output is a variable DC.

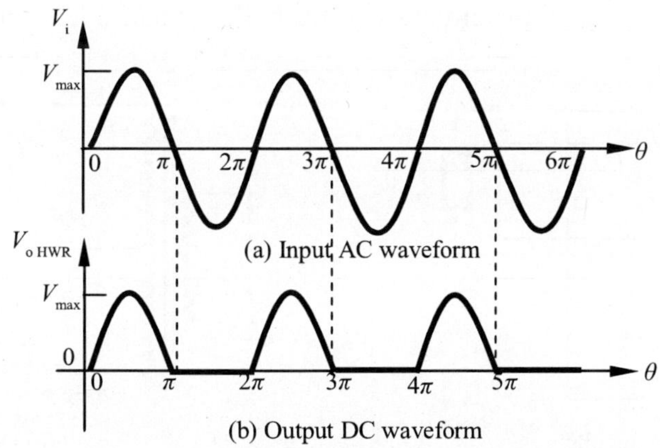

Fig. 1.21 Waveforms associated with HWR.

The conduction periods of the diode are also shown in Fig. 1.21(a). As can be seen, the diode will be ON during the intervals 0 to π, 2π to 3π, etc., and during these periods there will be current flow in the circuit. From Fig. 1.21(b), we observe that the diode remains OFF during the intervals π to 2π, 3π to 5π, etc., and during these periods there will be no current flow in the circuit.

1.9.2 Full-Wave Rectifier

In the full-wave rectifier (FWR), we rectify both the half-cycles of the AC input wave. For this purpose, we use either a centre-tapped transformer (CTT) and two diodes, as shown in Fig. 1.22(a), or four diodes and a transformer with no centre tap, as shown in Fig. 1.22(b).

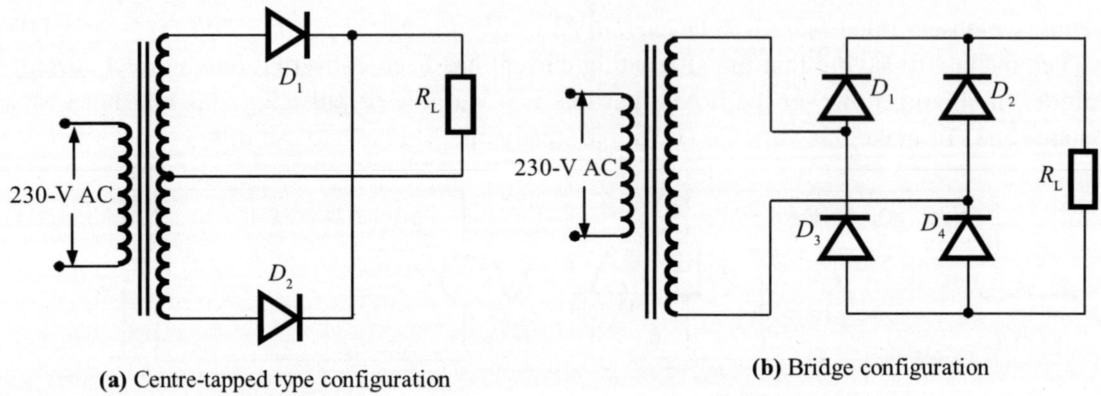

Fig. 1.22 Full-wave rectifier configurations.

Consider first the FWR with CTT. From the equivalent circuit for this condition shown in Fig. 1.23(a), we observe that during the positive half-cycles, D_1 conducts and D_2 remains off. Current I_{LP} flows through R_L in the direction shown. During the negative half-cycles, D_2 conducts and D_1 remains off as shown in Fig. 1.23(b), and we find that the diode current I_{LN} flows in the *same direction* through R_L. Thus this circuit rectifies both the half-cycles, and hence we call it the full-wave rectifier.

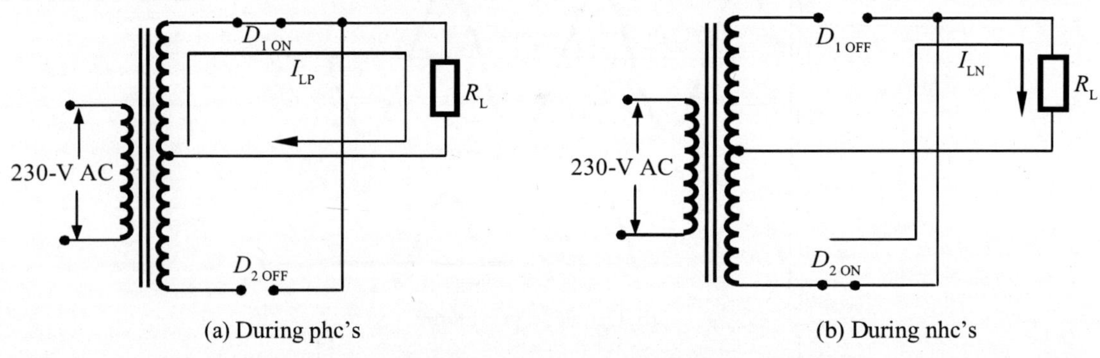

Fig. 1.23 Equivalent circuit of CT FWR during phc's and nhc's.

Consider next the FW bridge rectifier. As can be seen from the equivalent circuit for this condition shown in Fig. 1.24(a), during the positive half-cycles, diodes D_1 and D_4 conduct and D_2 and D_3 remain OFF. Current I_{LP} flows through R_L in the direction shown in the figure. During the negative half-cycles, D_2 and D_3 conduct and D_1 and D_4 remain OFF as shown in Fig. 1.24(b), and we find that the diode current I_{LN} flows in the *same direction* through R_L. Thus circuit rectifies both the half-cycles, and hence this is also a full-wave rectifier.

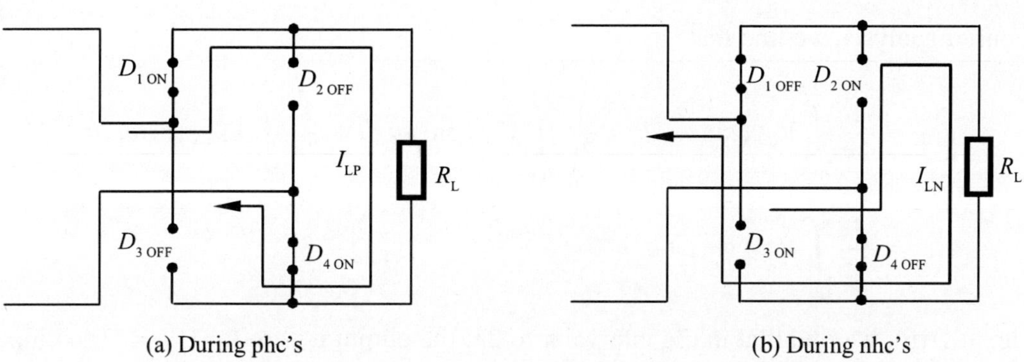

Fig. 1.24 Equivalent circuit of bridge WR during phc's and nhc's.

The waveforms associated with the full-wave rectifier are shown in Fig. 1.25. In the figure, we find that both the half-cycles are rectified. We can prove that by full-wave rectification, the output DC, and the rectification efficiency have been doubled, and the ripple reduced greatly. For further smoothening of the DC, we use appropriate filters in this case also.

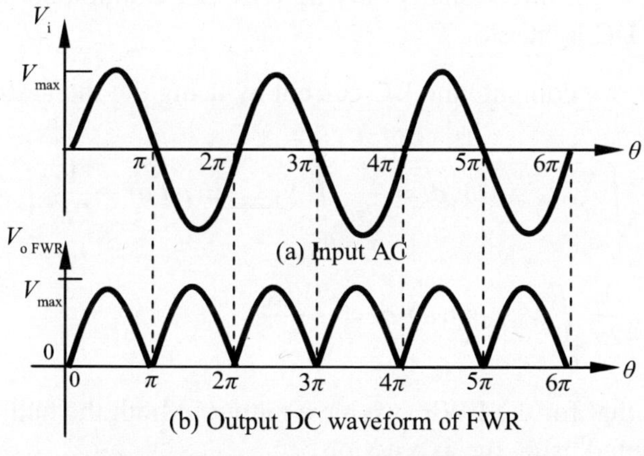

Fig. 1.25 Waveforms associated with FWR.

1.9.3 Analysis of the Half-Wave and Full-Wave Rectifiers

By analysis of the rectifier, we mean the determination of DC and RMS values of the output current and voltage. Through these analyses, we also determine the efficiency of the rectifier and the ripple content in the rectified waveform.

(a) Determination of the DC voltage and current

Let the input voltage be given by

$$v_i = V_{max} \sin \omega t \tag{1.8}$$

Using Fourier analysis, we find that

$$V_{DC} = \frac{1}{2\pi}\int_0^{2\pi} V_{max}\sin\omega t\,dt = \frac{1}{2\pi}\int_0^{\pi} V_{max}\sin\omega t\,dt + \frac{1}{2\pi}\int_{\pi}^{2\pi} V_{max}\sin\omega t\,dt$$

$$= \frac{1}{2\pi}\int_0^{\pi} V_{max}\sin\omega t\,dt + 0 = \frac{V_{max}}{\pi} \qquad (1.9)$$

From Fig. 1.21(b), we find that in the interval π to 2π, the output is zero for HWR. The output DC current can be obtained from the equation

$$I_{DC(HWR)} = V_{max}/\pi\,(r_t + r_d + R_L) \qquad (1.10)$$

where r_t = DC resistance of winding of the transformer used, and r_d = DC resistance of the diode used.

Note: The usage "DC current" is a wrong one in the strictest sense. This usage only means that the current is a *steady-state*, or uni-directionally varying one. For example, it is now quite common to use the term DC voltage, DC light, etc.

For the full-wave rectifier, we compute the DC current by using the integration

$$V_{DC(FWR)} = \frac{1}{2\pi}\int_0^{2\pi} V_{max}\sin\omega t\,dt = \frac{1}{2\pi}\int_0^{\pi} V_{max}\sin\omega t\,dt + \frac{1}{2\pi}\int_{\pi}^{2\pi} V_{max}\sin\omega t\,dt$$

$$= 2\times\frac{1}{2\pi}\int_0^{\pi} V_{max}\sin\omega t\,dt = \frac{2V_{max}}{\pi} \qquad (1.11)$$

Notice from Fig. 1.25(b) that for the FWR, we have output in both the half-cycles. The output DC current can now be computed using the expression

$$I_{DC(FWR)} = 2\,V_{max}/\pi\,(r_t + r_d + R_L) \qquad (1.12)$$

(b) Determination of the root mean-squared (rms) voltage and current

Mean-squared value of the voltage for the half-wave rectifier is given by the expression

$$V_{rms}^2 = \frac{1}{2\pi}\int_0^{2\pi} V_{max}^2\sin^2\omega t\,dt = \frac{1}{2\pi}\int_0^{\pi} V_{max}^2\sin^2\omega t\,dt = \frac{V_{max}^2}{4} \qquad (1.13)$$

From Eq. (1.13), taking the square root, we get

$$V_{rms(HWR)} = V_{max}/2 \qquad (1.14)$$

For the full-wave rectifier, we modify Eq. (1.13) as

$$V^2_{\text{rms(FWR)}} = 2 \times \left(V^2_{\max}/4\right) = V^2_{\max}/2 \tag{1.15}$$

From Eq. (1.15), we get

$$V_{\text{rms(FWR)}} = V_{\max}/\sqrt{2} \tag{1.16}$$

(c) Determination of the ripple factor

Ripple factor γ is a term associated with the amount of AC in the rectified DC voltage and current. For a pure DC, γ will be zero. The larger the value of γ, the more the AC content in the rectified DC. We define the ripple factor technically as

$$\gamma = v_{\text{ac}}/V_{\text{DC}} \tag{1.17}$$

where v_{ac} = AC content in the rectified voltage. From Fourier series expansion, we know that

$$V^2_{\text{rms}} = V^2_{\text{DC}} + v^2_{\text{ac}} \tag{1.18}$$

From Eqs. (1.17) and (1.18), we get

$$\gamma = \sqrt{\left(v^2_{\text{rms}}/V^2_{\text{DC}}\right) - 1} \tag{1.19}$$

Substituting respective values in Eq. (1.19), we get for half-wave rectifier

$$\gamma_{\text{HWR}} = \sqrt{\left(v^2_{\text{rms}}/V^2_{\text{DC}}\right) - 1} = \sqrt{\left[(V_{\max}/2)^2/(V_{\max}/\pi)^2\right] - 1} = 1.21 \tag{1.20}$$

For the full-wave rectifier, we obtain

$$\gamma_{\text{FWR}} = \sqrt{\left[\left(V_{\max}/\sqrt{2}\right)^2/(2V_{\max}/\pi)^2\right] - 1} = 0.48 \tag{1.21}$$

From Eqs. (1.20) and (1.21), we find that the ripple-content in the full-wave rectifier is much less than that in the half-wave rectifier, which indicates that the DC output of the full-wave rectifier is more steady than the DC output of the half-wave rectifier.

(d) Rectification efficiency

Rectifiers are AC-to-DC converters. We define rectification efficiency η *as the ratio of the DC power output to the AC power input*. In equation form, we write

$$\eta = \frac{\text{DC power output}}{\text{AC power input}} \times 100\% = \frac{P_{\text{O(DC)}}}{P_{\text{in(ac)}}} \times 100\% \tag{1.22}$$

For the half-wave rectifier, we have

$$P_{\text{O(DC)}} = I^2_{\text{DC}} R_{\text{L}} = I^2_{\max} R_{\text{L}}/\pi^2 \tag{1.23}$$

and

$$P_{\text{in(ac)}} = I^2_{\text{rms}}\,(r_t + r_d + R_L) = I^2_{\text{max}}\,(r_t + r_d + R_L)/4 \tag{1.23a}$$

Substituting for $P_{O(DC)}$ and $P_{\text{in(ac)}}$ into Eq. (1.22), we get

$$\eta_{\text{HWR}} = \frac{\left(I^2_{\text{max}}/\pi^2\right) R_L}{I^2_{\text{max}}/4\,(r_t + r_d + R_L)} \times 100 \tag{1.24}$$

From Eq. (1.24), we have

$$\eta_{\text{HWR}} = [R_L/(r_t + r_d + R_L)\,] \times 40.6\% \tag{1.25}$$

Equation (1.25) states that a maximum of about 40.6% of the AC input power is converted into DC output power by the half-wave rectifier.

For the full-wave rectifier, we find

$$\eta_{\text{FWR}} = \frac{\left(4I^2_{\text{max}}/\pi^2\right) R_L}{I^2_{\text{max}}/2\,(r_t + r_d + R_L)} \times 100 = \frac{R_L}{r_t + r_d + R_L} \times 81.2\% \tag{1.26}$$

We thus find that the efficiency of the full-wave rectifier is double that of the half-wave rectifier.

(e) Comparison of Full-Wave and Half-Wave Rectifiers

A comparison of the performances of the half-wave and full-wave rectifiers are shown in Table 1.2. In this table, we have used a parameter called PIV. This term stands for the peak-inverse voltage, which is defined as the *maximum reverse voltage that can be applied across a diode before it breaks down*. It may be noted that diodes with lower PIV costs less than those with higher PIV rating. Hence we find that, even though the bridge rectifier has to use four diodes as compared to the centre-tapped rectifier, it uses diodes with lower PIV's, which leads to lower cost. Also, the bridge rectifier uses a transformer that requires no centre taps. This is an important advantage because it is very difficult to get a transformer with the centre tap exactly in the middle of its winding.

Table 1.2 Comparison of rectifiers

Sl.No.	Parameter	HWR	FWR-CT	FWR-bridge
1.	$V_{O(DC)}$	V_{max}/π	$2V_{max}/\pi$	$2V_{max}/\pi$
2.	$V_{O(rms)}$	$V_{max}/2$	$V_{max}/\sqrt{2}$	$V_{max}/\sqrt{2}$
3.	PIV	V_{max}	$2V_{max}$	V_{max}
4.	γ	21	0.48	0.48
5.	η_{max}	40.6%	81.2%	81.2%
6.	No. of diodes	1	2	4

Yet another advantage of the bridge rectifier is that since current flows through the transformer in both the directions (upwards and downwards), transformer core saturation is avoided.

This saves power as compared to the centre-tapped rectifier. So, we conclude that the bridge rectifier is the most commonly used rectifier circuit among all the types of rectifiers.

1.10 RECTIFIERS WITH CAPACITOR FILTER

1.10.1 Half-Wave Rectifier with Capacitor Filter

Consider Fig. 1.26(a) shows the circuit of a HWR with a capacitor C_L connected in shunt with the load resistor R_L. During the phc's, the diode will conduct and charge the capacitor to the peak value of the input voltage, as shown in Fig. 1.26(b). Once the input falls below the peak value, the diodes become open circuited and the capacitor discharges through the load resistor, as shown in Fig. 1.26(b). It can be seen that the charging and discharging reduces the ripple content in the output waveform and smoothens the output to a more steady DC voltage. It may also be noted that the output is a perfect DC if $C_L = R_L = \infty$ so that the charge stored in the capacitor will not discharge and the output remains at a constant value.

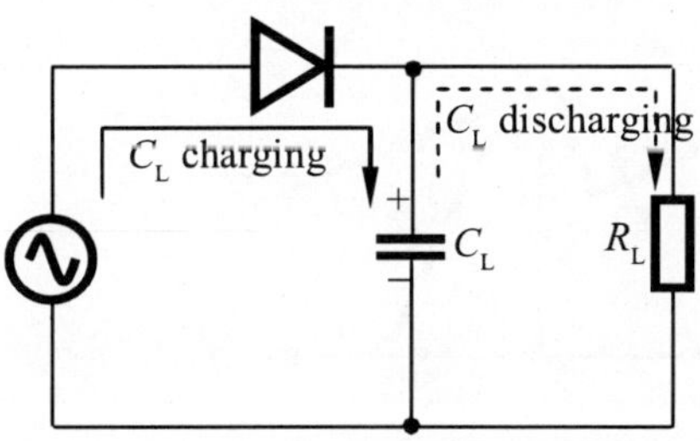

Fig. 1.26(a) Set-up of HWR with *C*-filter.

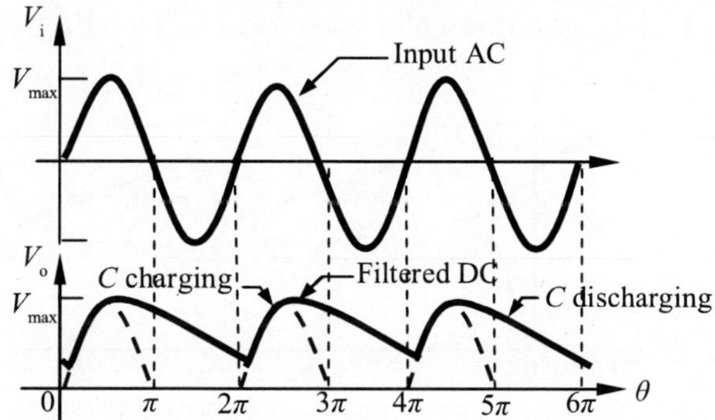

Fig. 1.26(b) Waveforms of HWR with *C* filter.

1.10.2 Full-Wave Rectifier with Capacitor Filter

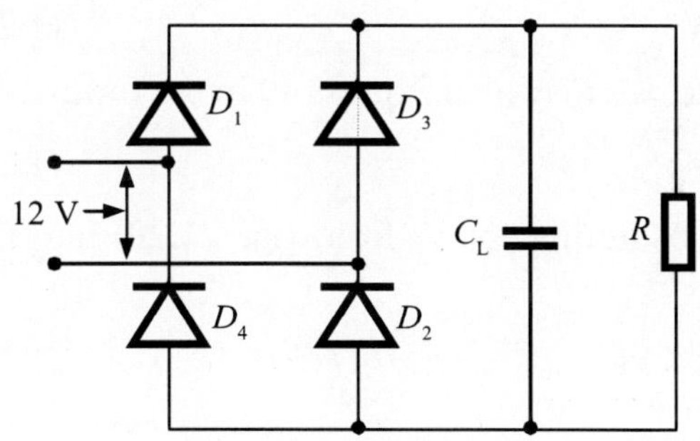

Fig. 1.27(a) Bridge rectifier with *C* filter.

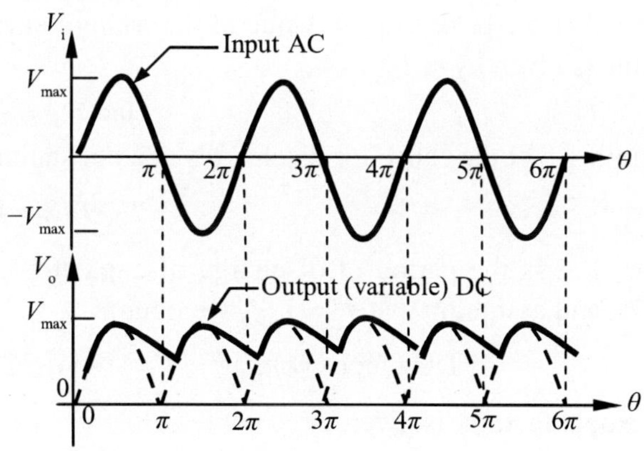

Fig. 1.27(b) Waveforms of FWR with *C* filter.

Consider now Fig. 1.27(a) shows the circuit of a FWR with a capacitor C_L connected in shunt with the load resistor R_L. During the phc's, diodes D_1 and D_2 will conduct, as shown in Fig. 1.27(b). During the nhc's, diodes D_3 and D_4 will conduct. It can be seen that during the phc's and nhc's, the capacitor gets charged to the peak value of the input voltage. Once the input falls below the peak value, the diodes become open circuited and the capacitor discharges through the load resistor, as shown in Fig. 1.27(b). It can be seen that the charging and discharging reduces the ripple content in the output waveform and smoothens the output to a more steady DC voltage than that obtained from the HWR with C filter. It may also be noted that the output is a perfect DC, if $C_L = R_L = \infty$ so that the charge stored in the capacitor will not discharge and the output remains at a constant value.

1.10.3 Approximate Analysis of Full-Wave Rectifier with Capacitor Filter

For approximate analysis, we make use of Fig. 1.28, which shows the output waveform shown in Fig. 1.27(b) as a triangular waveform. This will greatly simplify the computation.

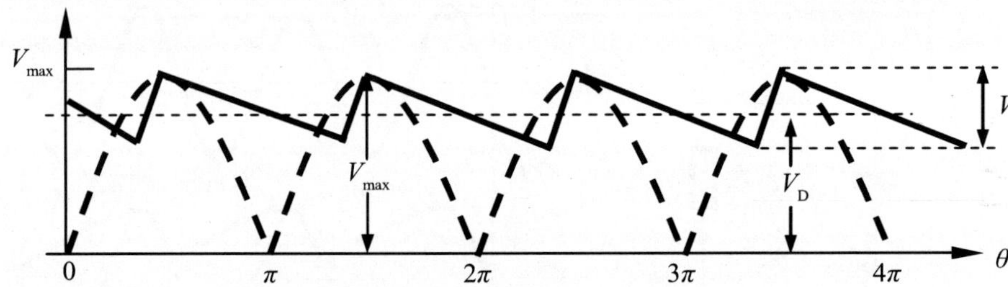

Fig. 1.28 Approximate triangular output waveform.

From the waveform shown, we make the following assumptions.

$$\text{DC output voltage} = V_{max} - (V_r/2) \tag{1.27}$$

where V_r is the peak-to-peak value of the triangular ripple voltage waveform. The rms value of this voltage is given by

$$V_{rms} = V_r/2\sqrt{3} \tag{1.28}$$

Assuming that the capacitor discharges at a constant rate, we may also express V_r in the form

$$I_{DC}T/2 = CV_r \tag{1.29}$$

where $T/2$ is the period of discharge of capacitor. Substituting for V_r from Eq. (1.29) into Eq. (1.27), and assuming that $T = 1/f$, we obtain

$$\text{DC output voltage} = V_{max} - (I_{DC}T/4C) = V_{max} - (I_{DC}/4fC) \tag{1.30}$$

The ripple factor γ is given by

$$\gamma = \frac{V_r}{V_{DC}} = \frac{I_{DC}}{4\sqrt{3}fCV_{DC}} = \frac{1}{4\sqrt{3}fCR_L} \tag{1.31}$$

where $V_{DC}/I_{DC} = R_L$, the load resistance.

1.11 APPLICATION OF ZENER DIODES

We have already seen that Zener diodes are semiconductor diodes that exhibit a constant reverse breakdown voltage. This characteristic is made use of in voltage regulating circuits.

1.11.1 Simple Zener Voltage Regulator

Figure 1.29 shows the circuit of a simple voltage regulator using Zener diode. As shown in the figure, we connect a Zener diode D_Z in parallel to a load resistance R_L and a variable DC source voltage $V_{in(var)}$. A resistance R_S is connected in series with the source and the load.

The source supplies a current I_S, which gets divided as the Zener current I_Z and the load current I_L, as shown in Fig. 1.29. In equation form, the current relation may be expressed as

$$I_S = I_Z + I_L \tag{1.32}$$

Let us assume that the output voltage increases. This increase occurs due to an increase in the input voltage or the load current. We know that the voltage across the Zener diode will remain constant between the maximum and minimum values of the Zener current. This means that the voltage across the load resistance connected in parallel to the Zener diode will have to remain constant at the Zener breakdown value, i.e., V_O remains regulated and will be a constant value.

To keep the Zener-diode voltage constant at V_Z, we find that the Zener current has to increase and the increased-load current has to decrease so that the total current from the source I_S always remains constant. Figure 1.30 shows the *V-I* characteristic of the Zener diode. It can be seen from the characteristic, as stated before, V_Z remains constant only in between $I_{Z\,max}$ and $I_{Z\,min}$. It may be noted that below $I_{Z\,min}$, the Zener diode will not function at all as a constant-voltage device; above $I_{Z\,max}$, it will get destroyed due to excessive current. So, Zener diodes are useful as voltage regulators only if they are operated in between $I_{Z\,min}$ and $I_{Z\,max}$.

The series resistance R_S is used to drop the difference in the voltage between the input V_{IN} and the Zener voltage V_Z, i.e.,

$$V_{RS} = V_{in(var)} - V_Z = I_S R_S \tag{1.33}$$

This means that, if the input voltage increases, then the excess voltage will be dropped across R_S since V_Z is constant. Thus R_S helps to regulate the variations in the input voltage. It may be noted that the input must be greater than V_Z for the Zener diode to produce regulation, but less than the value that causes the destruction of the Zener diode.

The reverse curve shows a breakdown point at $V_R = V_Z$. This voltage is called the Zener voltage. Even though V_Z is assumed to be a constant, in practical cases, it shows a tolerance of ± 10%. This means that if $V_Z = 10$ V, then this value can vary from 9 V to 11 V. From Fig. 1.30, we also notice that V_Z remains at a constant value in between $I_{Z\,min}$, and $I_{Z\,max}$. If the reverse current $I_R < I_{Z\,min}$, there will be no Zener or avalanche action; above $I_{Z\,max}$, the diode current will exceed the safe limits to destroy the diode.

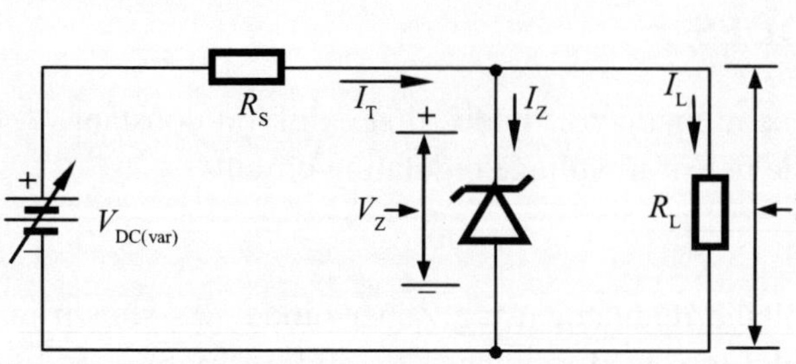

Fig. 1.29 Simple Zener-voltage regulator.

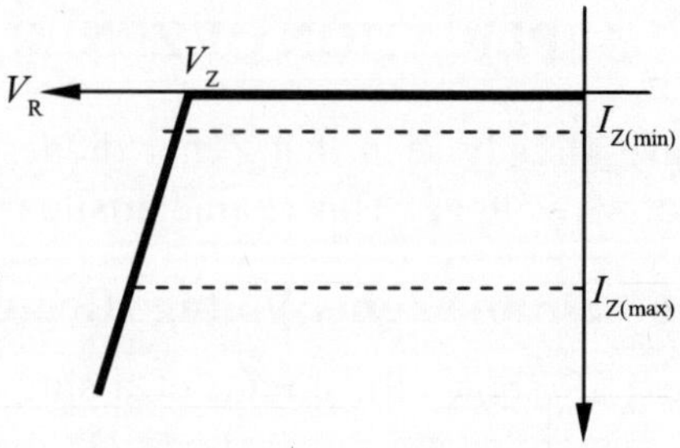

Fig. 1.30 Zener-diode characteristic.

Even though both the Zener and the avalanche diodes are different, it is quite common to call both the diodes by the common name Zener diode. Zener diodes find application in voltage regulators, and voltage references where the voltage should remain constant.

SOLVED EXAMPLES

S1.1 Compute the value of the current through the diode shown in Fig. S1.1(a).

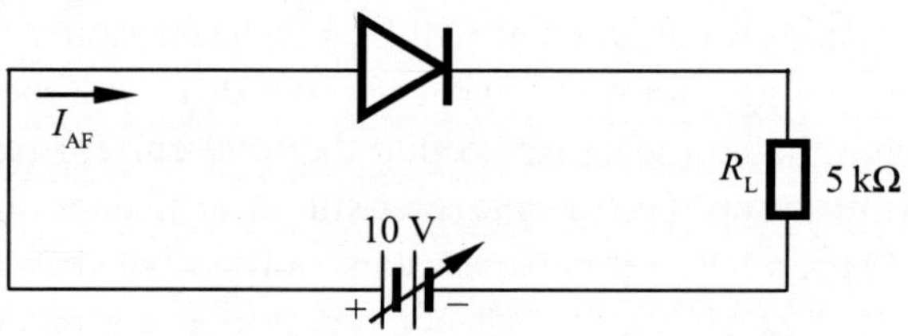

Fig. S1.1(a)

Solution

We have to determine the current assuming the diode as (a) ideal and (b) practical.

Case 1: Ideal diode condition: From Fig. S1.1(a), we find the current through the ideal diode

$$\mathbf{I_{AF} = V_{AA}/R_L = 10/5k = 2\ mA}$$

Case 2: Practical diode condition 1:

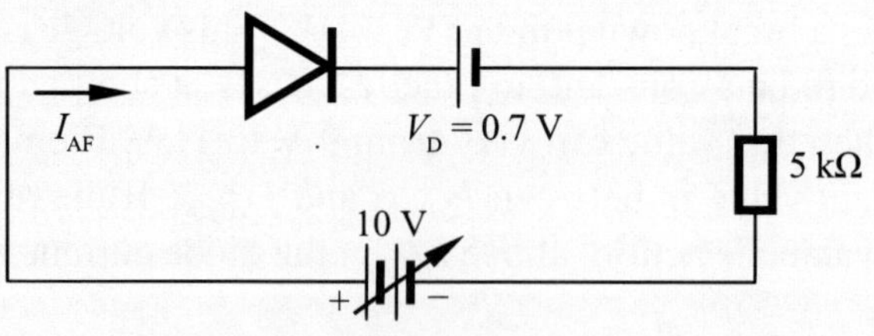

Fig. S1.1(b)

From Fig. S1.1(b), we assume that the diode cut-in voltage is present, but the forward dynamic resistance is absent in the system. We then can find the current through the diode (assuming built-in potential to be 0.7 V) as

$$\mathbf{I_{AF} = (V_{AA} - V_D)/R_L = (10 - 0.7)/5k = 1.86\ mA}$$

Case 3: Practical diode condition 2: From Fig. S1.1(c), we assume that the diode cut-in voltage and the dynamic forward resistance are present in the system. We then find the current through the diode (assuming built-in potential to be 0.7 V and resistance as 100 Ω) as

$$\mathbf{I_{AF} = (V_{AA} - V_D)/(r_f + R_L) = (10 - 0.7)/(5000 + 100) = 1.823\ mA}$$

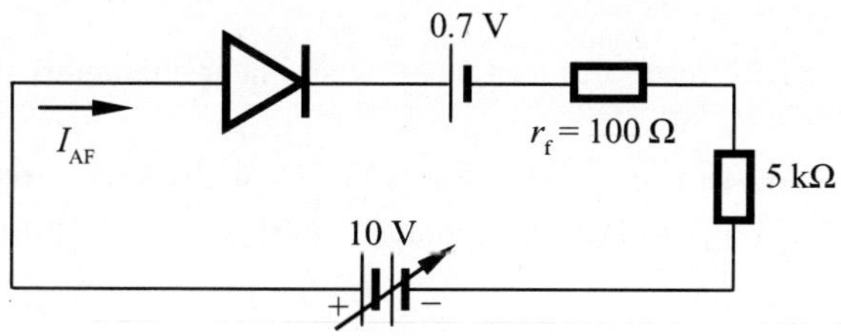

Fig. S1.1(c)

S1.2 Obtain the piece-wise linear characteristic of a silicon diode with $V_D = 0.7$ V, $r_f = 10\ \Omega$, and $I_{AF\,max} = 50$ mA.

Solution

From the given data, we find that the cut-in voltage of the diode is 0.7 V; up to this value current is zero in a piece-wise linear model. From 0.7 V the current rises with a slope of 10 Ω to a maximum value of 50 mA. Using these data, the piece-wise linear characteristic of the diode is constructed as shown in Fig. 1.2.

From the given data, we find that

$$r_f = \Delta V_F/\Delta I_F = 10\ \Omega \tag{1}$$

From Eq. (1), we obtain that

$$\Delta V_F = \Delta I_F \times r_f = (50\ \text{mA}) \times (10\ \Omega) \tag{2}$$

The starting point of the piece-wise linear characteristic is 0.7 V. Its maximum voltage-end value will be that corresponding to the maximum current multiplied the resistance plus the cut-in voltage of the diode. That is

$$\mathbf{V_{Fmax} = V_D + I_{Fmax} \times r_f = 0.7 + (50\ mA \times 10\ \Omega) = 0.7 + 0.5\ V = 1.2\ V} \tag{3}$$

Using the data obtained (i.e., cut-in voltage is 0.7 V, $V_{Fmax} = 1.2$ V, $I_{Fmax} = 50$ mA), we construct the piece-wise linear characteristic of the given silicon diode as shown in Fig. S1.2.

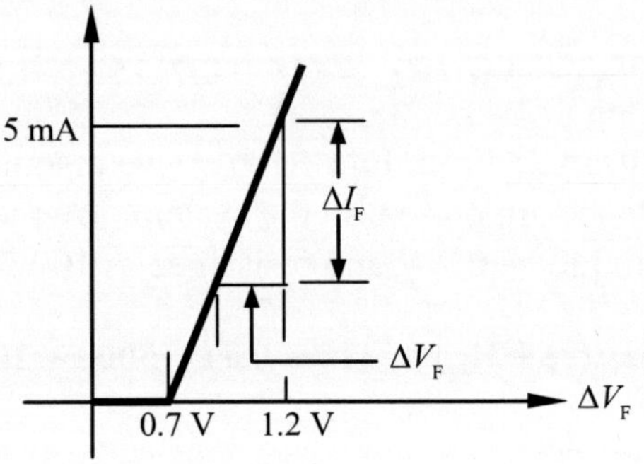

Fig. S1.2 Construction of piece-wise linear characteristic.

S1.3 A silicon diode has a de-rating factor of 6 mW/°C at 20°C. Compute the value of forward current I_F at 20°C and 60°C, if $V_D = 0.7$ V (constant) and the maximum power dissipation of the diode at 20°C is 500 mW.

Solution

The de-rating curve is shown in Fig. S1.3.

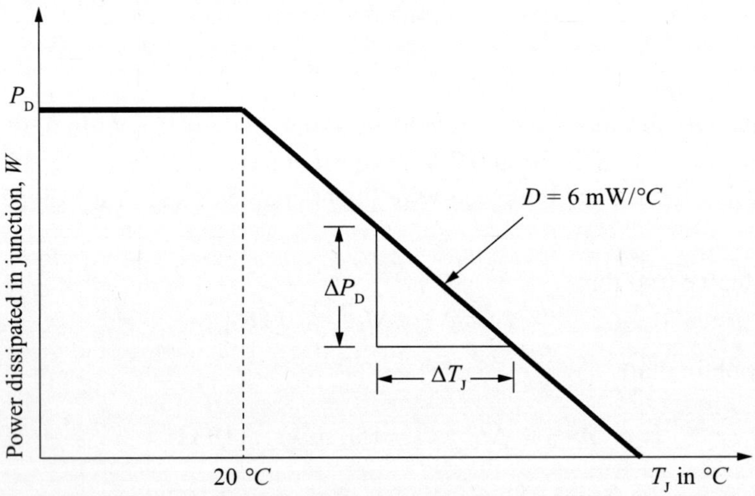

Fig. S1.3 Temperature de-rating curve of a diode.

We have the de-rating factor given by [see Eq. (1.7)]

$$D = \Delta P_D / \Delta T_j \text{ W/°C} = 6 \text{ mW/°C at } 20\text{°C} \tag{1}$$

From Fig. S1.3, we find

$$\Delta T_J = 60 - 20 = 40°C \quad (2)$$

Substituting for ΔT_J from Eq. (2) into Eq. (1), we get the change in the power dissipated at 60°C as

$$\Delta P_{D(60)} = 6 \times 10^{-3} \times 40 = 240 \text{ mW} \quad (3)$$

Since the maximum power dissipation capacity of the diode at 20°C is only 500 mW, we obtain the power dissipated at 60°C as

$$\mathbf{P_{D(60)} = 500 - 240 = 260 \text{ mW}} \quad (4)$$

The forward current for this power dissipation is

$$\mathbf{I_F = \frac{P_D}{V_F} = \frac{260}{0.7} = 371 \text{ mA}} \quad (5)$$

S1.4 A silicon diode has to be operated at constant forward current up to a temperature of 60°C. If V_D = 0.65 V at 30°C, calculate the value of V_D at 60°C. Also, compute the dynamic forward resistance at 30°C and at 60°C, if the forward current is 2 mA.

Solution

We have the voltage drop at any temperature given by the expression

$$V_{D(\text{any temp.})} = V_D \text{ at } T_1 + (\Delta T) \times (\Delta V_D/°C) \quad (1)$$

where from experiments it is found that $\Delta V_D/°C$ = −1.8 mV/°C for a silicon diode and −2.02 mV/°C for a germanium diode. Using the value of silicon diode in Eq. (1), we get

$$\mathbf{V_{D(60°)} = 0.65 + (60 - 30) \times \left(-1.8 \times 10^{-3}\right) = 0.596V} \quad (2)$$

The dynamic forward resistance r_f at any temperature is given by

$$r_f = \frac{V_T}{I_F}\left(\frac{T + 273}{273 + 30}\right) \quad (3)$$

where V_T is the thermal voltage given by kT/e = 0.025 mV/°C. Substituting given values in Eq. (3), we obtain the resistance at 30°C

$$\mathbf{r_{f(30)} = \frac{V_T}{I_F}\left(\frac{30 + 273}{273 + 30}\right) = \frac{0.025}{2 \text{ mA}} = 12.5\ \Omega} \quad (4)$$

The resistance at 60°C is given by

$$\mathbf{r_{f(30)} = \frac{V_T}{I_F}\left(\frac{60 + 273}{273 + 30}\right) = \frac{0.025}{2 \text{ mA}} \times \frac{333}{303} = 13.74\ \Omega} \quad (5)$$

S1.5 A certain Zener diode has a temperature coefficient of Zener voltage α_Z = 0.022%/°C. Compute the maximum current that may be allowed through that diode at 60°C, if V_Z = 5.1 V and the de-rating factor D = 5 mW/°C at 25°C, and power dissipation P_D = 500 mW at 25°C.

Solution

The Zener current at 25°C is given by

$$I_Z = \frac{P_D}{V_Z} = \frac{500 \times 10^{-3}}{5.1} = 98.04 \text{ mA} \quad (1)$$

The power dissipation at 60°C may be obtained using the expression

$$P_{D(60)} = P_{D(30)} - (\Delta T \times D) \quad (2)$$

Substituting values in Eq. (2) yields

$$P_{D(60)} = 500 \text{ mW} - [(60 - 25) \times 5 \text{ mW/°C}] = 325 \text{ mW} \quad (3)$$

The Zener voltage at any temperature is given by

$$V_{Z(\text{any temp.})} = V_Z \text{ at } T_1 + [(\Delta T) \times (\alpha_Z V_Z/100)] \quad (4)$$

Using Eq. (4), Zener voltage at 60°C is given by

$$V_{Z(60)} = 5.1 - [(60 - 25) \times (-0.022 \times 5.1/100)] = 5.061 \text{ V} \quad (5)$$

The maximum current at 60°C is now given by

$$\mathbf{I_Z = \frac{P_D}{V_{Z(60)}} = \frac{325 \times 10^{-3}}{5.061} = 64.22 \text{ mA}}$$

S1.6 A certain Zener diode has $I_Z = 100$ mA and $V_Z = 7.5$ V. Compute the maximum and minimum values of when I_Z changes by ±10 mA. The Zener-diode resistance $R_Z = 10\ \Omega$.

Solution

Since the change in $I_Z = 100$ mA is ±10 mA, we get

$$\Delta V_Z = \Delta I_Z \times R_Z = (\pm 10 \text{ mA})\,(10\ \Omega) = \pm 0.1 \text{ V}$$

Using the above values, we obtain the variations in V_Z as

$$V_{Z\min} = 7.5 - 0.1 = 7.4 \text{ V}$$

$$V_{Z\max} = 7.5 + 0.1 = 7.6 \text{ V}$$

S1.7 A certain Zener diode has an applied input voltage of 32 V. The series resistance is of 1 kΩ and the Zener voltage is $V_Z = 8.2$ V. Compute the Zener current, neglecting the Zener-diode resistance.

Solution

The Zener current is given by

$$\mathbf{I_Z = \frac{32 - 8.2}{1000} = 23.8\ mA}$$

S1.8 A 230-V, 50-Hz supply is connected to a half-wave rectifier. Compute the DC output voltage, and DC output current, if the load resistance is 1 kΩ.

Solution

The output DC voltage is given by

$$\mathbf{V_{DC(OUT)} = \frac{V_m}{\pi} = \frac{\sqrt{2}V_{rms}}{\pi} = \frac{\sqrt{2} \times 230}{\pi} = 103.54\ V}$$

The DC output current is given by

$$\mathbf{I_{DC(OUT)} = \frac{V_{DC(OUT)}}{R_L} = \frac{103.54V}{1000\Omega} = 103.54\ mA}$$

S1.9 A 230-V, 50-Hz supply is connected to a half-wave rectifier. Compute the maximum value of the output voltage, and maximum value of the output current, if the load resistance is 1 kΩ. Neglect the diode drop. Find out the peak inverse voltage also.

Solution

The maximum output voltage is given by

$$\mathbf{V_{OUTmax} = \sqrt{2} \times V_{rms} = \sqrt{2} \times 230 = 325.27\ V}$$

The peak-inverse voltage is given by

$$\mathbf{PIV = 325.27\ V}$$

The maximum output current is given by

$$\mathbf{I_{OUTmax} = \frac{V_{OUTmax}}{R_L} = \frac{325.27V}{1000} = 325.27\ mA}$$

S1.10 Compute value of the load current if the peak-to-peak ripple voltage is 0.5 V and the capacitor filter used is of 100 μF value. Find the average value of the output voltage also, if $R_L = 10$ kΩ.

Solution

We have the peak-to-peak ripple voltage given by

$$V_R = \frac{I_L T}{C} = I_L \times \left(\frac{1}{50}\right)\left(\frac{1}{100 \times 10^{-6}}\right) = 0.5 \text{ V (value given)}$$

The load current is then given by

$$\mathbf{I_L = \frac{V_R \times C}{T} = \frac{0.5 \times 100 \times 10^{-6}}{1/50} = 2.5 \text{ mA}}$$

The average output voltage is

$$\mathbf{V_{OUT(av)} = I_L \times R_L = 2.5 \times 10^{-3} \times 10 \text{ k}\Omega = 25 \text{ V}}$$

S1.11 Compute the DC and RMS values of output voltage and current in the case of a half-wave rectifier constructed using 1N 4001 diode. The input is 230-VAC. The diode DC resistance $r_D = 10\ \Omega$ and the transformer used is a 230 V/12 V step-down one having a DC resistance $r_t = 100\ \Omega$. The load resistance $R_L = 1000\ \Omega$. Compute also the ripple factor and efficiency of the rectifier.

Solution

From analysis, we know that

$$\mathbf{V_{OUT(DC)} = \frac{V_m}{\pi} = \frac{\sqrt{2}\, V_{RMS}}{\pi} = \frac{\sqrt{2} \times 12}{\pi} = 5.4 \text{ V}}$$

$$\mathbf{V_{OUT(RMS)} = \frac{V_m}{2} = \frac{\sqrt{2}\, V_{RMS}}{2} = \frac{\sqrt{2} \times 12}{2} = 8.49 \text{ V}}$$

$$\mathbf{I_{OUT(DC)} = \frac{V_{OUT(DC)}}{r_t + r_D + R_L} = \frac{5.4 \text{ V}}{100 + 10 + 1000} = 4.86 \text{ mA}}$$

$$\mathbf{I_{OUT(RMS)} = \frac{8.49\text{V}}{100 + 10 + 1000} = 7.65 \text{ mA}}$$

The ripple factor is given by

$$\mathbf{r = \sqrt{\frac{V_{RMS}^2 - V_{DC}^2}{V_{DC}^2}} = \sqrt{\frac{8.49^2 - 5.4^2}{5.4^2}} = 1.21}$$

The rectification efficiency is given by

$$\mathbf{\eta = \frac{P_{ODC}}{P_{INAC}} = \frac{I_{DC}^2 R_L}{I_{RMS}^2 (r_t + r_D = R_L)} = \frac{4.86^2 \times 1000}{7.65^2 \times 1110} \times 100 = 36.36\%}$$

S1.12 Compute the above parameters, if the rectifier used is a full-wave one.

Solution

From analysis, we know that

$$V_{OUT(DC)} = \frac{2V_m}{\pi} = \frac{2 \times \sqrt{2}V_{RMS}}{\pi} = \frac{2 \times \sqrt{2} \times 12}{\pi} = 10.8\ V$$

$$V_{OUT(RMS)} = \frac{V_m}{\sqrt{2}} = \frac{\sqrt{2}\,V_{RMS}}{\sqrt{2}} = 12\ V$$

$$I_{OUT(DC)} = \frac{V_{OUT(DC)}}{r_t + r_D + R_L} = \frac{2 \times 5.4\ V}{100 + 10 + 1000} = 9.72\ mA$$

$$I_{OUT(RMS)} = \frac{12\ V}{100 + 10 + 1000} = 10.81\ mA$$

The ripple factor is given by

$$r = \sqrt{\frac{V_{RMS}^2 - V_{DC}^2}{V_{DC}^2}} = \sqrt{\frac{12^2 - 10.8^2}{10.8^2}} = 0.484$$

The rectification efficiency is given by

$$\eta = \frac{P_{ODC}}{P_{INAC}} = \frac{I_{DC}^2 R_L}{I_{RMS}^2 (r_t + r_D = R_L)} = \frac{9.72^2 \times 1000}{12^2 \times 1110} \times 100 = 59.11\%$$

ONE-WORD/MULTIPLE-CHOICE QUESTIONS

M1.1 A point-contact diode is characterised by a metallic spring anode called ________.

M1.2 The junction potential of a point-contact diode is about ________.

M1.3 The junction potential of a Si Schottky-barrier diode is about ________.

M1.4 The reverse leakage current in a point-contact diode is ________.

M1.5 Depletion layer is called so because it is depleted of ________ charge carriers.

M1.6 We can measure the junction potential of a diode with a voltmeter ________.

M1.7 Surface-leakage current obeys ________ law.

M1.8 The expression for drift-current density is ________.

M1.9 The expression for conductivity is ________.

M1.10 The expression is $J = \sigma E$ is a statement of the ________ law.

M1.11 The expression for the electron diffusion current is ________.

M1.12 Avalanche breakdown is also called ________ ionisation.

M1.13 Zener breakdown is also called ________ ionisation.

M1.14 Zener breakdown occurs in ________ doped diodes.

M1.15 Avalanche breakdown occurs in ________ doped diodes.

M1.16 Reverse-saturation current due to EHPs is a constant because the number of EHPs for a given ________ is a constant.

M1.17 The expression for the hole diffusion current is ________.

M1.18 The DC output voltage of a half-wave rectifier is given by ________.

M1.19 The DC output voltage of a full-wave centre-tapped rectifier is given by ________.

M1.20 The DC output voltage of a full-wave bridge rectifier is given by ________.

M1.21 The rms output voltage of a full-wave centre-tapped rectifier is given by ________.

M1.22 The rms output voltage of a full-wave bridge rectifier is given by ________.

M1.23 The rms output voltage of a half-wave rectifier is given by ________.

M1.24 The maximum rectification efficiency of a full-wave centre-tapped rectifier is ________.

M1.25 The maximum rectification efficiency of a full-wave bridge rectifier is ________.

M1.26 The maximum rectification efficiency of a half-wave rectifier is given by ________.

M1.27 The ripple factor of a full-wave centre-tapped rectifier is ________.

M1.28 The ripple factor of a full-wave bridge rectifier is ________.

M1.29 The ripple factor of a half-wave rectifier is given by ________.

M1.30 The PIV of the diodes in a full-wave centre-tapped rectifier is ________.

M1.31 The PIV of the diodes of a full-wave bridge rectifier is ________.

M1.32 The PIV of the diode of a half-wave rectifier is given by ________.

REVIEW QUESTIONS

1. Explain structure of a conventional *PN* diode.
2. Describe the condition of a *PN* diode immediately after formation.
3. What happens in a *PN* diode under forward bias?
4. What happens under reverse bias?
5. Explain the *VI* characteristics of a *PN* diode.
6. What is drift current?
7. Derive an expression for drift current?
8. Derive the expressions for mobility of electrons and holes in *PN* diode.
9. Obtain an expression for the total current in a *PN* diode.
10. Draw the energy-level diagram of a *PN* diode.
11. Describe the avalanche breakdown.
12. Explain Zener breakdown.
13. Explain the behaviour of Zener and avalanche diodes based on *VI* characteristics.
14. Describe the constructions of Zener and Avalanche diodes.
15. Compare Zener and Avalanche breakdowns.
16. Describe the with suitable diagram, the working of a simple Zener-voltage regulator.
17. Draw the circuit diagram of half-wave rectifier and explain its working.
18. With suitable diagram, explain the working of a centre-tapped full-wave rectifier.
19. Draw the circuit diagram of a bridge rectifier and explain its operation.
20. Derive the expressions of V_{dc}, I_{dc}, V_{rms}, I_{rms}, efficiency, and ripple factor of

    a. Half-wave rectifier b. Full wave rectifier
21. Compare half-wave and full-wave rectifiers.
22. Compare full-wave centre-tapped and full-wave bridge rectifiers.
23. Explain the working of capacitor-input filter with reference to half-wave rectifier.

24. Explain the working of capacitor-input filter with reference to full-wave rectifier.

25. Explain the working of an *LC* filter.

26. Prove that the ripple factor of a half-wave rectifier is 1.21 and full-wave rectifier is 0.48.

27. Show that the maximum efficiency of half-wave rectifier is 40.6%.

28. Explain why a bridge rectifier is preferred to a centre-tapped rectifier. Is there any application where a centre-tapped rectifier is preferred over a bridge rectifier?

29. Draw the output waveform of a half-wave rectifier. Show the modification in the waveform when a capacitor is connected across the load resistor of this rectifier.

UNSOLVED PROBLEMS

P1.1 For an abrupt junction, the junction potential is 0.6 V. If the depletion width is 5 μm, compute the value of the proportionality constant between the two values.

[**Hint:** use the relation $k_j = W_j/\sqrt{V_j}$]

P1.2 For an abrupt junction, the junction potential is 0.6 V. If the depletion width is 4 μm, compute the value of the proportionality constant between the two values.

P1.3 For a linearly graded junction, the junction potential is 0.3 V. If the depletion width is 2 μm, compute the value of the proportionality constant between the two values.

[**Hint:** use the relation $k_j = W_j/V_j^{1/3}$]

P1.4 For a linearly graded junction, the junction potential is 0.6 V. If the depletion width is 3 μm, compute the value of the proportionality constant between the two values.

P1.5 For a linearly graded junction, find the junction potential, if the depletion width is 3.56 μm and the value of the proportionality constant between the two is 4.0 μm/V$^{1/3}$.

P1.6 Compute the value of electron drift current density J_{ndrift} in a given semiconductor material, if the number of electrons in one cm^3 of that material is $2 \times 2 \times 10^{16}$. The mobility of electrons in that material is 1000 cm^2/V-s, and the applied electric field intensity is 15 V/cm.

[**Hint:** use the relation $J_{ndrift} = ne\mu_n E$]

P1.7 Compute the value of electron drift current density J_{ndrift} in a given semiconductor material, if the number of electrons in one m^3 of that material is 2×10^{22}. The mobility of electrons in that material is 100 m^2/V-s, and the applied electric field intensity is 150 V/m.

P1.8 Compute the value of conductivity of a given semiconductor material, if the number of electrons in one cm^3 of that material is 1×10^{16}. The mobility of electrons in that material is 1300 cm^2/V-s, and the applied electric field intensity is 10 V/cm.

[**Hint:** use the relation $\sigma = ne\mu_n$]

P1.9 Compute the value of resistivity of a given semiconductor material, if the number of electrons in one cm^3 of that material is 1×10^{16}. The mobility of electrons in that material is 1300 cm^2/V-s, and the applied electric field intensity is 10 V/cm.

[**Hint:** use the relation $\rho = 1/\sigma = 1/ne\mu_n$]

P1.10 Compute the value of mobility of electrons in a given semiconductor material, if the number of electrons in one m^3 of that material is 2×10^{16}. The current density in the material is 50 A/cm^2 and the applied electric field intensity is 15 V/cm.

[**Hint:** use the relation $J_{ndrift} = ne\mu_n E$]

P1.11 A half-wave rectifier is used to convert 230-V AC into DC across a load of 1 kΩ. The transformer used is of the 230-V/12-V type. The DC resistance of the transformer used is 12 Ω, and the resistance of the diode used is 22 Ω. Compute:

(a) The DC output voltage
(b) The rms value of the output voltage
(c) The Ripple factor
(d) The Rectification efficiency

P1.12 A full-wave centre-tapped rectifier is used to convert 230-V AC into DC across a load of 1 kΩ. The transformer used is of the 230-V/12-0-12-V type. The DC resistance of the transformer used is 12 Ω, and the resistance of the diode used is 22 Ω. Compute:

(a) The DC output voltage
(b) The rms value of the output voltage
(c) The Ripple factor
(d) The Rectification efficiency

P1.13 A full-wave bridge rectifier is used to convert 230-V AC into DC across a load of 1 kΩ. The transformer used is of the 230-V/12-0-12-V type. The DC resistance of the transformer used is 12 Ω, and the resistance of the diode used is 22 Ω. Compute:

(a) The DC output voltage
(b) The rms value of the output voltage
(c) The Ripple factor
(d) The Rectification efficiency

P1.14 A full-wave rectifier is used to convert 230-V AC into DC across a load of 2.2 kΩ. The transformer used is of the 230-V/9-0-9-V type. The DC resistance of the transformer used is 10 Ω, and the resistance of the diode used is 50 Ω. Compute the following Compute the relevant parameters if the rectifier is connected in the (i) in the centre-tapped mode (ii) in the bridge mode:

(a) The DC output voltage
(b) The rms value of the output voltage
(c) The Ripple factor
(d) The Rectification efficiency
(e) Peak-inverse voltage (PIV) the diodes used
(f) Compare the results.

P1.15 A full-wave rectifier is used to convert 230-V AC into DC across a load of 100 Ω. The transformer used is of the 230-V/15-0-15-V type. The DC resistance of the transformer used is 10 Ω, and the resistance of the diode used is 100 Ω. Compute the following if the rectifier is connected in the (i) in the centre-tapped mode (ii) in the bridge mode and compare the results:

(a) The DC output voltage
(b) The rms value of the output voltage
(c) The Ripple factor
(d) The Rectification efficiency
(e) Peak-inverse voltage (PIV) the diodes used
(f) The rating of the load resistance used

CHAPTER 2

BIPOLAR JUNCTION TRANSISTORS

2.1 INTRODUCTION

Figure 2.1 shows the sandwiched model of a bipolar junction *PNP* transistor. It consists of a thin *N* region, sandwiched between two P^{++} regions. The comparatively thinner of the two P^{++} regions is called the **emitter**, and the other the **collector**. The thinnest central *N* region is called the **base** since it supports the collector and emitter regions on either side of it.

The word **transistor** was coined from the words **transfer** and **resistor**. The name transistor thus represents the transfer of a low input resistance to a high output resistance.

2.2 CONSTRUCTION OF TRANSISTORS

2.2.1 Alloy-Junction Transistor

Figure 2.2 shows the construction of an *alloy-junction* transistor, which was used before the planar technology for the construction of semiconductor devices was invented. An alloy-junction transistor is made up of a very thin piece of germanium/silicon wafer called the *base*, on either side of which two indium dots are alloyed (by heating) to form two alloy junctions, as shown. These junctions form the emitter and collector depletion layers in the base layer. It can be seen that this construction produces a $P^{++}NP^{++}$ transistor (where P^{++} represents heavy doping and *N* represents light doping). The whole assembly is finally encapsulated in a hermetically sealed metallic or plastic outer casing.

2.2.2 Planar Diffused Construction of Transistors

Figure 2.3 shows the construction of a modern planar diffused transistor. It consists of a lightly doped *N* substrate of silicon, which acts as the collector. Into this *N* substrate, we first diffuse *P*-type atoms (such as boron atoms) to neutralise the *N*-type atoms (such as phosphorous atoms); then to make this region *P*, we add more *P* atoms so that this region becomes medium doped (P^{+}) base. Theoretically, base must be lightly doped to reduce recombination that results in the reduction of collector current.

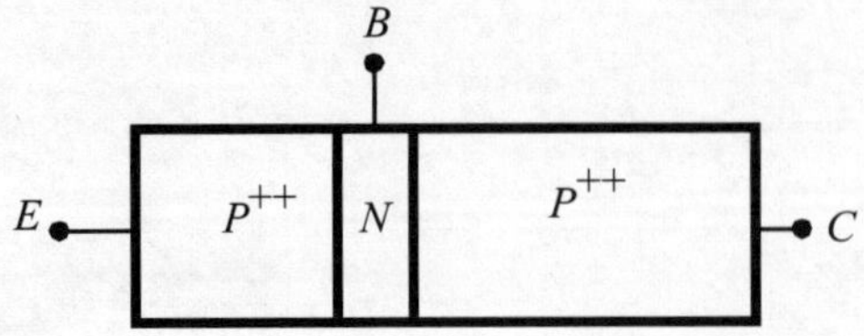

Fig. 2.1 Sandwiched model of an alloy transistor.

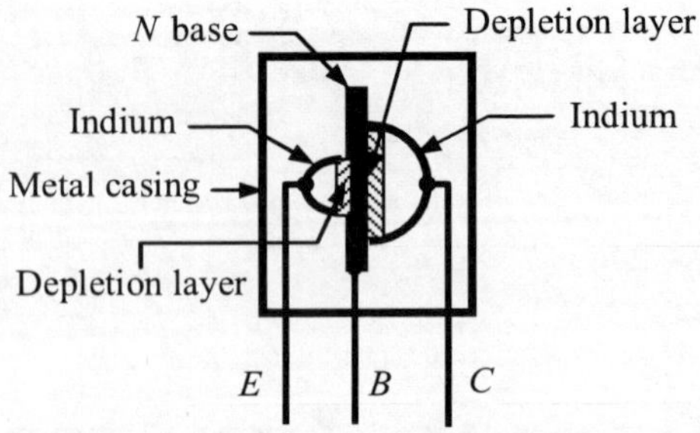

Fig. 2.2 Alloy-junction transistor.

In the planar construction, we are forced to make the base P^+, which increases the probability of recombination in the base. Fortunately, we can make the base very thin (as shown in Fig. 2.3) by diffusion, which will reduce recombination in the base.

Into the medium-doped P base, we diffuse more atoms of N to form a heavily doped (N^{++}) emitter region. It may be noted that emitter must be heavily doped as it has to supply a large number of electrons to produce the emitter-current flow. We now grow a thin layer of SiO_2 over the emitter region, as show in Fig. 2.3. This will protect the emitter surface from external contamination, and is known as *surface-passivation layer*. Windows are now cut at appropriate emitter and base points and aluminium metallization done to take out external contacts.

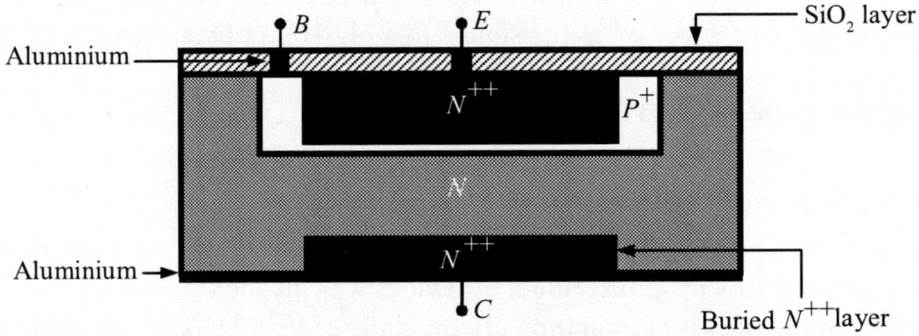

Fig. 2.3 Planar-diffused transistor.

There is a buried N^{++} layer at the bottom of the collector to prevent the formation of a Schottky-barrier diode between the collector N and the bottom aluminium deposition. The collector contact is then taken out from the aluminium deposition.

2.3 WORKING OF BIPOLAR JUNCTION TRANSISTORS

A transistor (bipolar junction transistor, as it is called now) may be thought of as two PN-junction diodes connected back-to-back. These are, respectively, the emitter-base and collector-base diodes.

For the transistor to perform amplification, its emitter-base (EB) junction is to be *forward-biased* and collector-base (CB) junction *reverse-biased*, as shown in Fig. 2.4. The configuration shown in Fig. 2.4 is called the *common-base configuration* (CBC), since the base is common for both the input and output sections.

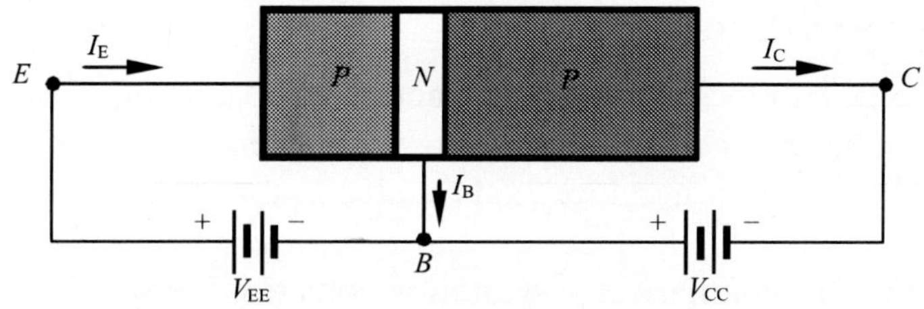

Fig. 2.4 Construction of an *NPN* transistor.

Now, consider the *E-B* junction. Under forward bias, holes are injected form the emitter into the base to produce the emitter current I_E. If the base is very thin, these holes will go to the collector.

The holes injected into the base however face the problem of recombination with the electrons in that region. However, with a narrow base, the chances of recombination are very small. The holes reaching the collector from the emitter are pulled out by the negative terminal of the battery to produce the collector current I_C. We notice that, with the collector connected to the negative of the battery, we have a reverse-bias existing in the *C-B* junction.

From the discussions given above, we conclude that, **for proper transistor operation, the *E-B* junction must be forward biased, and the *C-B* junction must be reverse biased**. The *resistance of a forward-biased diode is very low, and that of a reverse-biased diode is very high.* Hence the operation of the transistor may be regarded as the transfer of a low input resistance to a high output resistance, leading to the name **trans**fer res**istor**, or **transistor**.

If I_E represents the emitter current, I_B represents the base current, and I_C represents the collector current, we have the basic transistor current relation

$$I_E = I_C + I_B \tag{2.1}$$

2.4 COMMON-BASE CHARACTERISTICS OF BJT

Figure 2.5 shows the experimental set-up for plotting the input and output characteristics of a silicon transistor in the common-base characteristics.

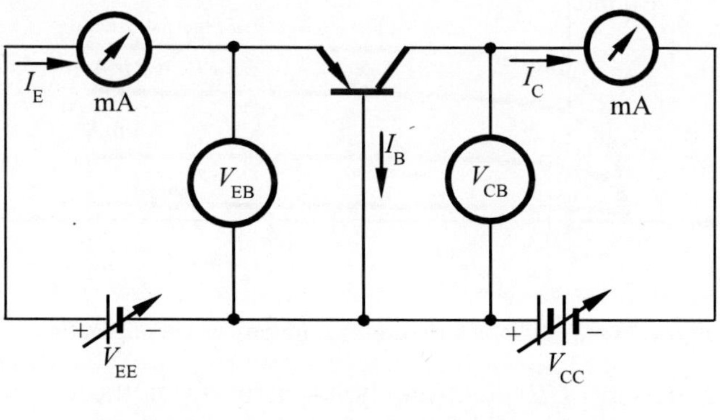

Fig. 2.5 Setup for plotting the *CB* characteristics.

2.4.1 Input Characteristics in CBC

Input characteristic of a BJT in the common-base mode is the variation of emitter current I_E with emitter-base voltage V_{EB}, keeping collector-base voltage V_{CB} constant. Figure 2.6 shows one typical input characteristic. The curve shown in Fig. 2.6 is similar to that of the forward characteristic of a *PN* junction diode. In the figure, we find that the cut-in voltage is above 0.4 V to 0.5 V depending on the construction of the BJT. Similarly, the saturation voltage is above 0.7 V to 0.8 V.

Above 0.8 V, I_E may rise to extremely high values so that the transistor may get destroyed. In between 0.4 to 0.8 V, the device operates in its **active region**, where I_E varies directly with respect to V_{EB}. We can see that when V_{CB} is increased, I_E also gets increased. This is due to what is called the **Early effect** [see Sub-section 2.4.2].

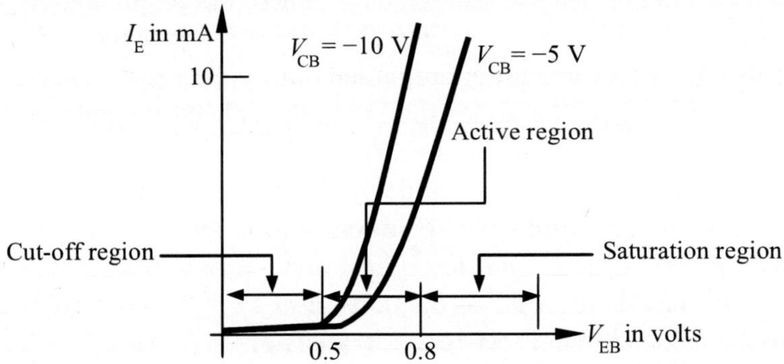

Fig. 2.6 Input characteristics in the common-base configuration.

2.4.2 Output Characteristics in CBC

Common-base output characteristic is the variation of collector current I_C with collector-to-base voltage V_{CB}, keeping emitter current I_E constant. Figure 2.7 shows the output characteristics in the common-base configuration for various values of I_E.

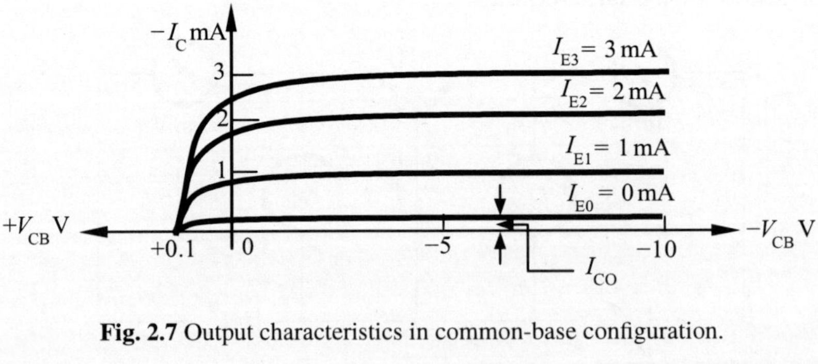

Fig. 2.7 Output characteristics in common-base configuration.

With reverse bias across the *CB* junction, holes arriving at the collector from the emitter-through-base are pulled out of the collector by the positive terminal of the battery to produce I_C. Initially, when the *EB* junction is open-circuited, no holes will be injected from the emitter into the base, and $I_E = 0$. However, with the emitter open, we find that a reverse saturation current I_{CO} flows between the base and the collector (*CB* acting as a reverse-biased diode). Usually, I_{CO} will be in the nano- or micro-ampere range. Even when V_{CB} is increased, I_C will remain constant at I_{CO}, because it is constant for a given temperature.

Now, fix I_E at 1 mA and vary V_{CB} from 0 volt to 10 volts in regular steps of 2 volt each and note the corresponding I_C. It can be observed that I_C is more or less constant and equal to I_E, as shown in Fig. 2.7. This is because all the holes injected from the emitter, except those recombined in the base, will arrive at the collector. Since the base is very thin, the number of holes lost due to recombination will be usually very small, so that I_C is almost equal to I_E. Thus we find that the collector current is *almost* constant.

The above statement is now slightly modified: collector current will increase slightly with collector-to-base voltage due to what is known as the *Early effect.* We know that width of the depletion region varies with applied voltage, i.e., with forward bias, it will decrease and with reverse bias, it will increase, as shown in Fig. 2.8. Since in a BJT, *EB* junction must always be forward biased and *CB* junction reverse biased, the base width will effectively get decreased with an increase in the collector-base voltage; this in turn reduces the recombination in the base and produces slight increase in the collector current with increase in V_{CB}.

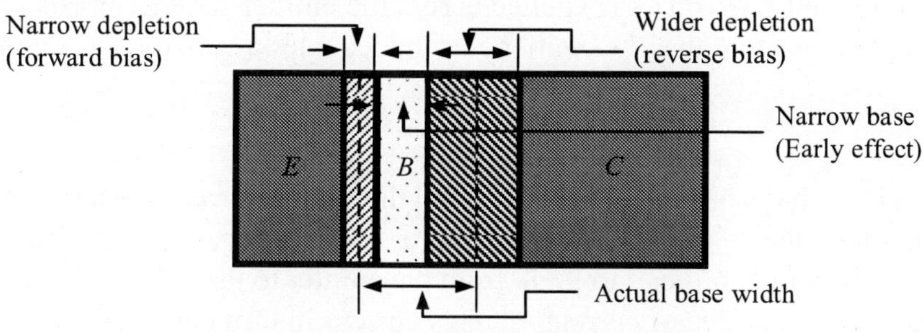

Fig. 2.8 Early effect (base-width modulation).

Inspection of the curves shown in Fig. 2.7 reveals that at $V_{CB} = 0$ also $I_C \approx I_E$. This is because the condition $V_{CB} = 0$ represents a shorted collector-to-base region; in this state, holes reaching collector will be pulled out by the *negative terminal* of the *emitter battery*, as shown in Fig. 2.9, which means that the collector current is still flowing and its value is almost equal to that of the emitter current.

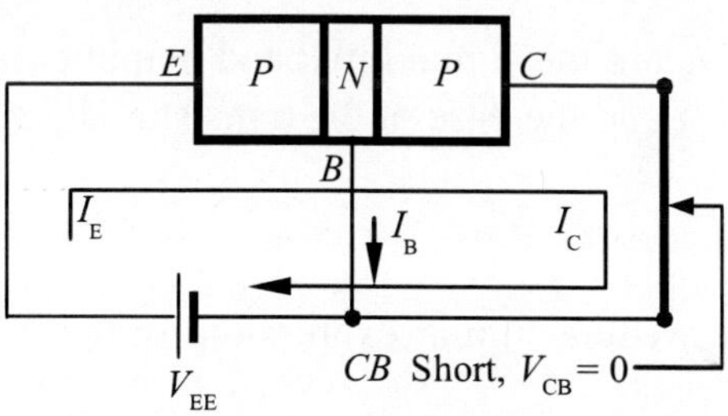

Fig. 2.9 Collector current at $V_{CB} = 0$.

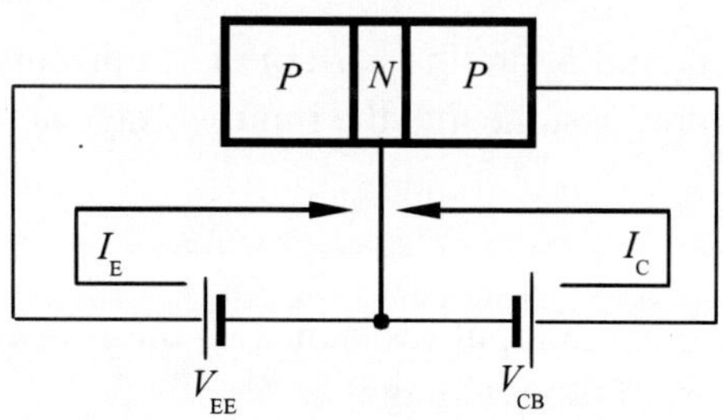

Fig. 2.10 Collector current when $V_{CB} = V_{EE}$.

Thus we find that $I_C = I_E$ even at $V_{CB} = 0$. Now to make $I_C = 0$, we find from Fig. 2.10 that the voltages V_{EE} and V_{CB} must be equal and opposite so that the currents I_C and I_E are also equal and opposite. This is called the saturation condition. From the characteristic curves shown in Fig. 2.7, we notice this region as that marked by $+V_{CB} = 0.1$ V. Thus **in saturation, both the *CB* and *EB* junctions are forward biased.**

We now define *the DC current amplification factor* in the common-base configuration

$$\alpha_{DC} = I_C/I_E \tag{2.2}$$

For a good BJT, I_C is almost equal to I_E. In typical cases, we observe that $I_C = 0.99$ to $0.999\ I_E$, so that α varies approximately between 0.99 and 0.999.

2.4.3 Common-Base Amplifier

Amplification is defined as the *process of controlling the flow of energy, the controlling energy being negligibly smaller than the controlled energy*. This is the ideal definition of an amplifier. This is now extended to prove that a bipolar junction transistor can perform amplification.

Consider a BJT connected in the *CB* configuration to a load resistance R_L as shown Fig. 2.11. The signal to be amplified $V_{max} \sin \omega t$ is applied across the emitter-base terminals of the transistor. With the signal applied, we find that the emitter current becomes a variable DC, represented by:

$$v_{EB} = V_{EB} + V_{max} \sin \omega t \tag{2.3}$$

where V_{EB} = DC bias voltage between the emitter and the base, and v_{EB} = instantaneous total base-emitter voltage (sum of the V_{EB} and $V_{max} \sin \omega t$). As can be seen, v_{EB}, which is a variable DC voltage, drives a variable DC emitter current i_E from the emitter to the collector, where it becomes a corresponding variable DC collector current i_C. This current in turn develops a variable DC output voltage v_L across the load resistance R_L, as shown in Fig. 2.11. The original signal $V_{max} \sin \omega t$ can be recovered from this output voltage v_L by capacitor coupling, which will remove the DC from v_L and deliver the AC signal output voltage v_o. From Fig. 2.11, we observe that

$$i_E = I_E + i_e \tag{2.4}$$

$$i_C = I_C + i_c \tag{2.5}$$

where i_E and i_C are the variable DC currents, and i_e and i_c are the signal input and output currents. Let us now assume that the input voltage across the emitter and the base of the transistor is given by

$$v_i = i_e \times r_i \tag{2.6}$$

where r_i is the input resistance of the transistor. We can write a similar expression for the voltage across R_L (output voltage) as

$$v_L = i_c \times R_L = v_o \tag{2.7}$$

We define voltage amplification of the transistor as

$$A_v = \frac{\text{output voltage}}{\text{input voltage}} = \frac{i_c R_L}{i_e r_i} = \frac{i_c}{i_e} \times \frac{R_L}{r_i} = \alpha_{ac} \frac{R_L}{r_i} \quad (2.8)$$

where the $\alpha_{ac} = i_c/i_e$ is the current amplification factor under signal condition in the common-base configuration. It may be noted that $\alpha_{ac} = \alpha_{dc}$ for all practical purposes. Its value may be taken to be approximately equal to 1. Then from Eq. (2.8), we find that, if $R_L > r_i$, *signal amplification has indeed taken place in the common-base configuration.*

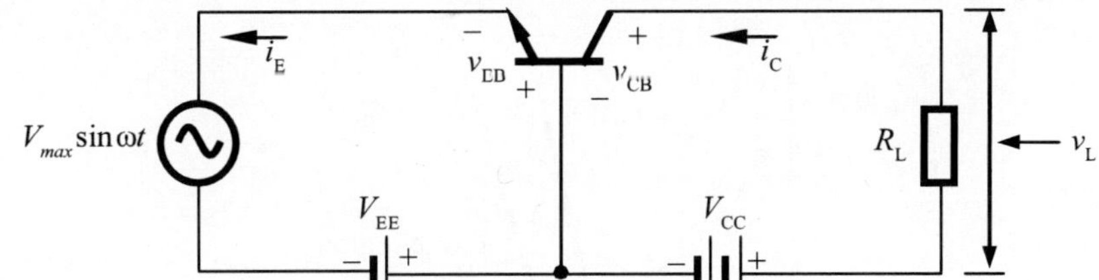

Fig. 2.11 BJT as a *CB* amplifier.

2.5 COMMON-EMITTER CONFIGURATION (CEC)

Figure 2.12 shows the common-emitter configuration (*CEC*). Here, the emitter is common for both the input and output sections. For the *CEC*, the base-emitter region is forward biased in the usual sense, i.e., the *P* base is connected to the positive terminal and the *N* emitter connected to the negative terminal of the battery. However, we observe that the reverse bias between collector and base is not direct. This is because battery terminals are connected to the collector and emitter terminals. For direct reverse bias, this must be between collector and base. In the *CEC*, the base is indirectly connected to the battery negative through the emitter region that acts as a conductor.

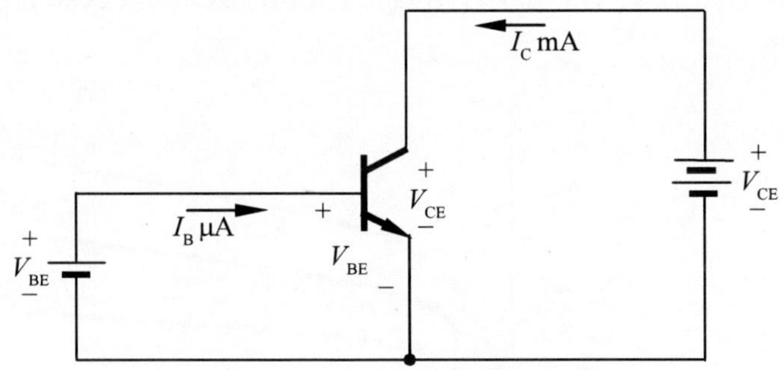

Fig. 2.12 BJT amplifier in *CEC*.

In the common-emitter configuration, the input current is I_B and the input voltage is V_{BE}. The output current and voltage are, respectively, I_C and V_{CE}.

2.5.1 Input Characteristics in CEC

The input characteristics (V_{BE}-I_B) of the *CEC* are shown in Fig. 2.13. We find that these curves resemble those of a forward-biased diode. The input current I_B is in microamperes and the output current I_C is in milliamperes. In Fig. 2.13, we find that V_{BEC} (some authors use the notation V_γ) represents the *cut-in* voltage of the transistor. V_{BEC} is usually around 0.4 to 0.5 V. Similarly, V_{BES} (some authors use the notation V_σ) represent the *saturation* value of the base-emitter voltage. Usual value of V_{BES} is around 0.7 to 0.8 V. In between the cut-in (or cut-off) and saturation values of V_{BE}, we have the *active region* of operation of the transistor.

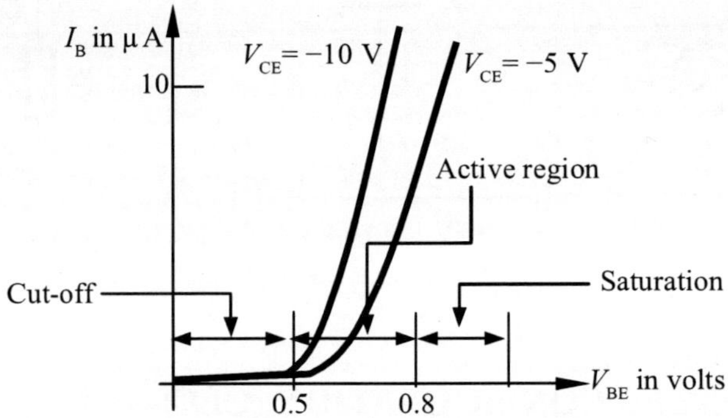

Fig. 2.13 Input characteristics in the common-base configuration.

2.5.2 Output Characteristics in CEC

Figure 2.14 shows the common-emitter output curves (V_{CE}-I_C) for various fixed values of I_B. It can be seen that, if we extend these characteristics backwards, they are seen to meet at a common point on the *x*-axis. This voltage, shown in Fig. 2.14, is the Early voltage V_A and is a constant (about 50 to 100) for a given BJT. The Early voltage is now used as a design parameter in modern computer-aided designs.

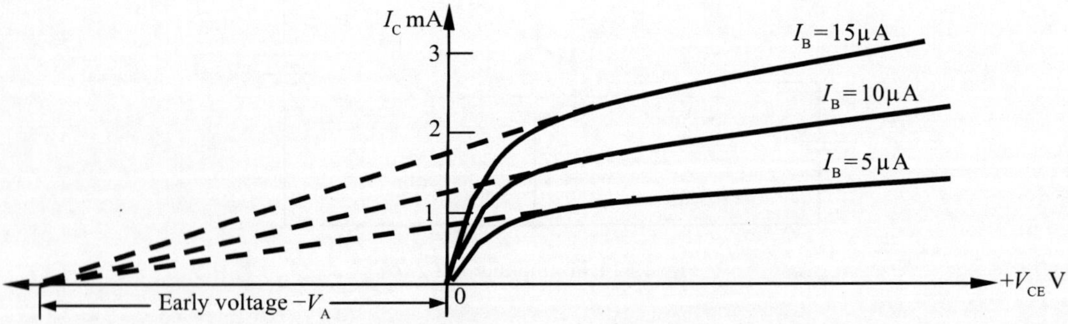

Fig. 2.14 Output characteristics in common-emitter configuration.

It can be seen that the maximum permissible swing in the input signal is between these two extreme values, and may be approximately calculated as $(0.8 - 0.5) = 0.3$ V (peak-to-peak). This means that, if we apply an AC input signal of value equal to $0.15 \sin \omega t$ to the base of the transistor, we will get an undistorted output across its collector.

From Fig. 2.14, we find that in the case of the *CEC* also, there are three operating regions, viz., cut-off, saturation, and active.

The saturation region is characterised by a low value of V_{CE} (0.1 to 0.2 V). In this condition both the *EB* and *CB* junctions are forward biased. The cut-off region is that region in the characteristic where $V_{BE} < 0.5$ V. In this case $I_C = 0$.

We find that the *saturation and cut-off are the stable states* of a transistor. In the cut-off state, $I_C = 0$, and in the saturation state, I_C is a maximum. A transistor having these operating states is a switch. Such transistor switches are widely used in modern digital circuits. It is quite important to note that transistors, operated in their stable states, do not require special temperature-stabilising circuits.

The region in the characteristics between cut-off and saturation is called the active region. In this region, the collector current I_C is directly proportional to the collector-emitter voltage V_{CE}. This region is used for the amplification of signals. A transistor operated in the active region is operating in an *unstable* condition. Therefore, a *transistor operated in its active region requires special stabilising circuits for its stable operation.*

2.5.3 Current Amplification Factor in CEC

We call the DC current-amplification factor in *CEC* as β_{DC} and define it as

$$\beta_{DC} = I_C / I_B \tag{2.9}$$

Since I_C is in milliamperes and I_{BC} is in microamperes, we find that there is a very large current amplification in *CEC*, compared to *CBC*. We may also define the AC current-amplification factor in *CEC* β_{ac}

$$\beta_{ac} = \frac{i_c}{i_b} = \frac{\Delta\, i_C}{\Delta\, i_B} \tag{2.10}$$

2.5.4 Relation between α and β

We may develop a relation between DC values of α and β as given below. From fundamental definitions, we have

$$\alpha = I_C / I_E \tag{2.11}$$

$$\beta = I_C / I_B \tag{2.12}$$

$$I_E = I_C + I_B \tag{2.13}$$

Substituting for I_E from Eq. (2.13) into Eq. (2.11), we have

$$\frac{I_C}{\alpha} = I_C + I_B \tag{2.14}$$

Rearranging Eq. (2.14) and solving, we get

$$\beta = \alpha/(1 - \alpha) \tag{2.15}$$

We can also prove the inverse relation

$$\alpha = \beta/(1 + \beta) \tag{2.16}$$

Usually α lies in the range of 0.99 to 0.999 so that β will lie in the range of 99 to 999. Since $\beta \gg 1$, we find that *common-emitter configuration* gives much more power amplification than *common-base configuration*. As a result of this, we find that the *common-emitter configuration has become the most commonly and widely used amplifier configuration all the BJT configurations.*

2.5.5 Amplification in Common-Emitter Configuration

Figure 2.16 shows an amplifier configuration using the *CEC*. Let the input voltage be given by:

$$v = V_{max} \sin \omega t \tag{2.17}$$

This input voltage is superimposed over the base-emitter bias V_{BE}, to give a variable DC input voltage

$$v_{BE} = V_{BE} + V_{max} \sin \omega t \tag{2.18}$$

This variable DC voltage will drive a variable DC input current i_B through the base-emitter path. This base current in turn will drive a variable DC collector current i_C through the load resistance R_L developing a variable DC output voltage $v_L = v_o$, given by:

$$v_o = i_C \times R_L \tag{2.19}$$

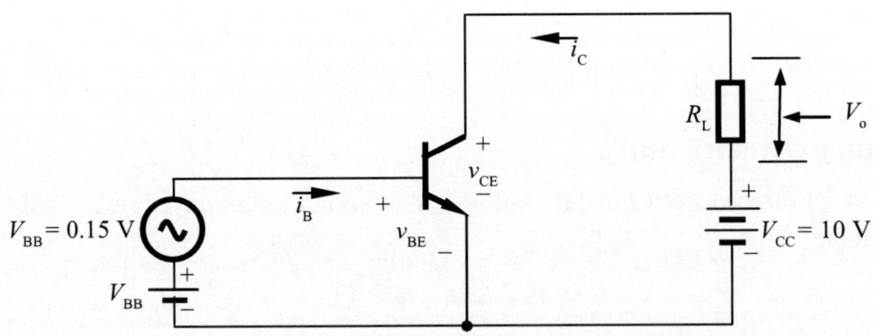

Fig. 2.15 BJT amplifier in *CEC*.

Assuming an input resistance of value r_i, we may write the actual input voltage to be applied at the input of the amplifier as:

$$v_i = i_B \times r_i \tag{2.20}$$

From Eqs. (2.19) and (2.20), we get voltage amplification in *CEC* as:

$$A_v = \frac{i_C \times R_L}{i_B \times r_i} = \beta \frac{R_L}{r_i} \tag{2.21}$$

Since $\beta \gg 1$, even if $R_L = r_i$, we find that this configuration gives very large voltage amplification.

2.6 COMMON-COLLECTOR CONFIGURATION (CCC)

In this case, collector is common to the input and output sections, as shown in Fig. 2.16. The biasing still follows the same rule: *EB junction forward biased, and CB junction reverse biased.* The respective biasings are as shown in Fig. 2.16. The input current is I_B and the output current is I_E. The current gain therefore is

$$\gamma = \frac{\boldsymbol{I_E}}{\boldsymbol{I_B}} = \frac{\boldsymbol{I_C + I_B}}{\boldsymbol{I_B}} = \boldsymbol{1 + \beta} \tag{2.22}$$

We find that the current gain γ in *CCC* is greater than the current gain β in *CEC* by 1. This means that *CCC* will produce more power amplification than *CEC*. However, we also find (and can prove) that voltage gain in *CCC* is *close to and less than unity.*

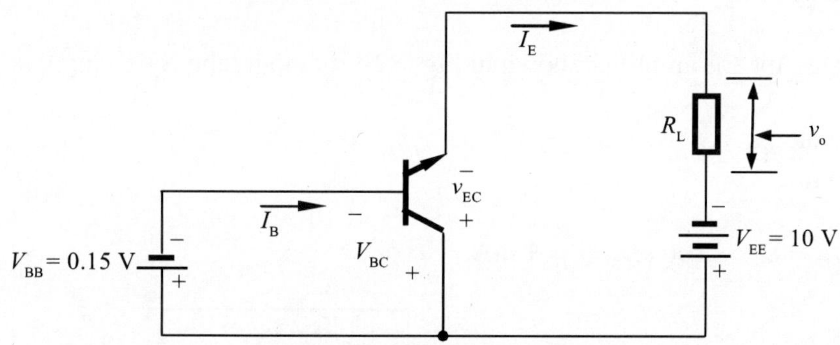

Fig. 2.16 BJT amplifier in *CCC*.

2.6.1 Input Characteristics of CCC

Figure 2.17 shows the input characteristics (V_{BC} vs I_B) of the *CCC* configuration. The characteristics are seen to be almost vertical, indicating that a small increase in the input voltage produces a large increase in the input current. This is expected because the input voltage (V_{CB}) is a reverse-bias (*CB* junction to be reverse-biased for proper transistor operation), and we know that a reverse-bias has very little influence over the flow of current.

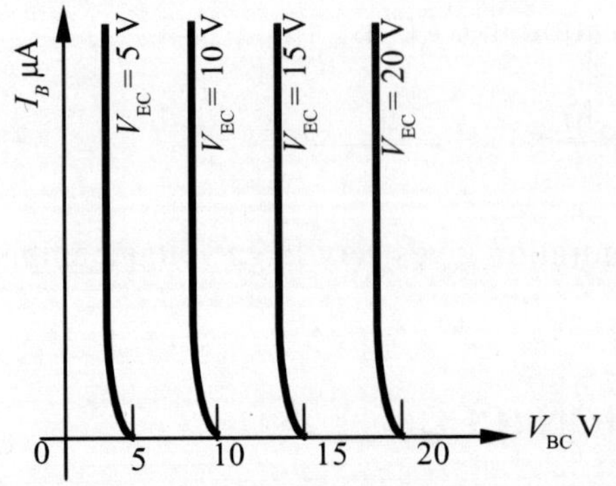

Fig. 2.17 Input characteristics in CCC.

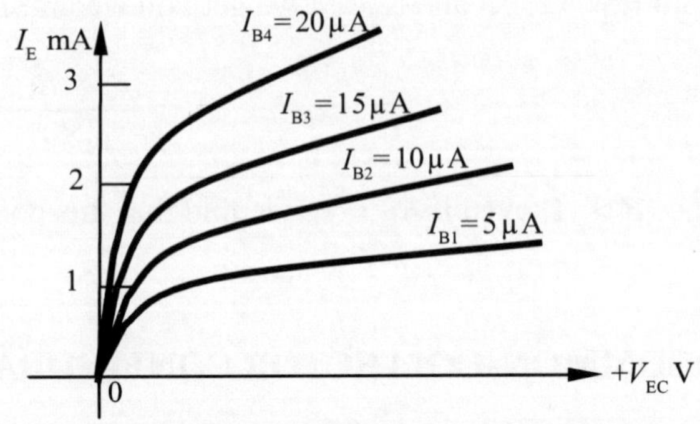

Fig. 2.18 Output characteristics in CCC.

2.6.2 Output Characteristics of CCC

Since the output is the collector-emitter region, the output characteristics will be similar to that of the *CE* output characteristics. The major difference between the two is that, in the case of *CCC*, the output current is I_E, whereas this is I_C in the case of *CEC*. Figure 2.18 shows the output characteristics in *CCC*, a plot between V_{EC} and I_E.

SOLVED EXAMPLES

S2.1 For the transistor *CE* amplifier shown in Fig. S2.1, compute the following:

(a) Collector voltage
(b) Voltage gain

Given that $\beta_{ac} = \beta_{dc} = 100$ and $I_C = 1$ mA.

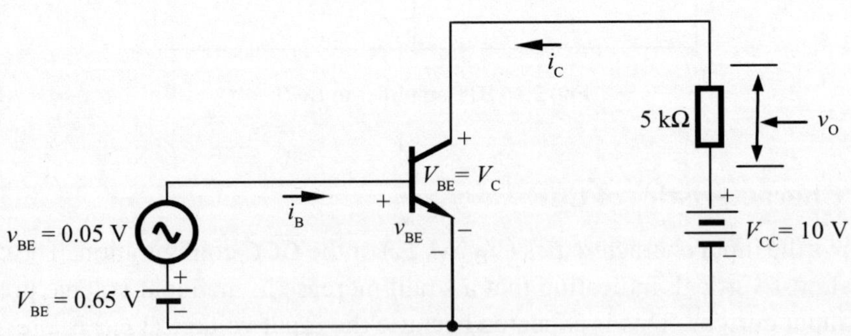

Fig. S2.1

Solution

From the given figure, we find that

$$V_{CE} = V_{CC} - I_C R_L$$

Substituting given values and solving, we obtain the collector voltage

$$\mathbf{V_C = V_{CE} = 10 - 1 \times 5 = 5\ V}$$

The base current for the DC state is

$$I_B = \frac{I_C}{\beta} = \frac{1\text{ mA}}{100} = 0.01\text{ mA}$$

Assuming saturation to be 0.8 V and cut-off to be 0.5 V, and $V_{BEQ} = 0.65$ V, we find the input voltage swing to be (0.8 – 0.5) = ± 0.15 V. Corresponding to this base-voltage swing, let the base-current swing be $\pm\Delta I_B$ (obtainable from a V_{BE}/I_B plot). Using this data, we find the output current swing to be

$$\Delta I_C = \beta \times \Delta I_B = 100\Delta I_B$$

Then output swing

$$\Delta V_{CE} = 100 \times \Delta I_B \times R_C = 5 \times 10^5 \Delta I_B$$

Let the base current swing be ±10 μA, then

$$\Delta V_{CE} = \left(5 \times 10^5\right) \times \left(\pm 10 \times 10^{-6}\right) = \pm 5\text{ V}$$

The voltage gain therefore is

$$\mathbf{A_v} = \frac{\pm\mathbf{5}}{\pm\mathbf{0.15}} = \mathbf{33.33}$$

S2.2 For the *CE* amplifier shown in Fig. S2.2, assume that the input is 3 V. Compute the values of I_C and I_B and β_{DC} of the transistor.

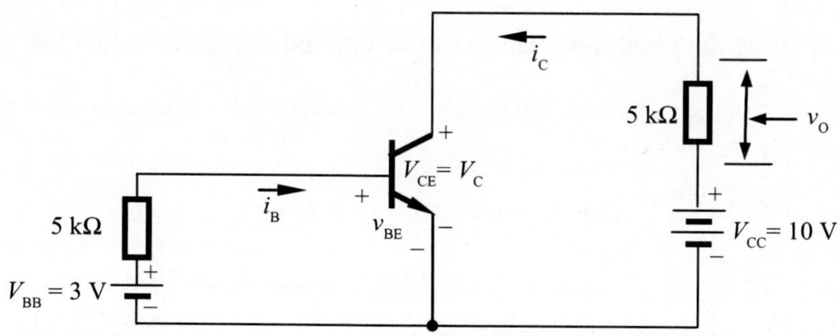

Fig. S2.2

Solution

Since the input voltage is greater than 8.8 V, we can assume that the transistor is in saturation. This means that V_{BESAT} = 0.8 V and V_{BESAT} = 0.2 V. Using these values, we obtain

$$\mathbf{I_{BSAT}} = \frac{\mathbf{V_{IN} - V_{BESAT}}}{\mathbf{R_B}} = \frac{\mathbf{3 - 0.8}}{\mathbf{22k}} = \mathbf{100\ \mu A}$$

$$\mathbf{I_{cSAT}} = \frac{\mathbf{V_{CC} - V_{CESAT}}}{\mathbf{R_C}} = \frac{\mathbf{10 - 0.2}}{\mathbf{5k}} = \mathbf{1.96\ mA}$$

The current gain under saturation is given by

$$\beta_{DC(SAT)} = \frac{\mathbf{I_{CSAT}}}{\mathbf{I_{BSAT}}} = \frac{\mathbf{1.96 \times 10^{-3}}}{\mathbf{100 \times 10^{-6}}} = \mathbf{19.6}$$

It can be seen that the gain of the transistor under saturation condition is much smaller than its value under active-region operation.

ONE-WORD/MULTIPLE-CHOICE QUESTIONS

M2.1 Amplification factor is defined as ________.

M2.2 Transconductance is defined as ________.

M2.3 Drain resistance is defined as ________.

M2.4 The channel enhanced in a MOSFET is known as ________ layer.

M2.5 The word *Q*-point stands for ________.

M2.6 Another name for operating point is ________-point.

M2.7 Operating point means ________ state condition.

M2.8 The line joining the maximum values of current and voltage is called the ________ line.

M2.9 The biasing arrangement that makes use of a fixed resistor between the collector and base of a transistor is called ________ bias.

M2.10 Collector current is given by the expression $I_C = \alpha I_E +$ ________.

M2.11 Collector current is given by the expression $I_C =$ ________ $+ I_{CO}$.

M2.12 I_C, V_{BE} and g_m are related by the expression ________.

M2.13 Transconductance of a BJT is defined as ________.

M2.14 Transconductance of a BJT may be approximately expressed in terms of V_T as ________.

M2.15 Current stability factor is defined as ________.

M2.16 Voltage stability factor is defined as ________.

M2.17 Current-gain stability factor is defined as ________.

M2.18 Current stability factor for the case of a voltage-divider biasing is given by ________.

M2.19 Voltage stability factor for the case of a voltage-divider biasing is given approximately by ________.

M2.20 The variation of collector current with temperature leading to the destruction of the transistor is known as ________.

M2.21 The diode biasing used in integrated circuits is known as ________.

M2.22 The biasing obtained by using a source resistance in JFETs is known as ________.

M2.23 The biasing obtained by using a voltage-divider circuit in JFETs is known as ________.

M2.24 The hybrid-π equivalent of h_{ie} is ________.

M2.25 The hybrid-π equivalent of h_{fe} is ________.

M2.26 The hybrid-π equivalent of h_{re} is ________.

M2.27 The hybrid-π equivalent of h_{oe} is ________.

REVIEW QUESTIONS

1. Explain how a bipolar junction transistor can act as an amplifier.
2. What do you mean by operating point?
3. What is load line?
4. What is meant by biasing of amplifier? Why do you require it?
5. Explain fixed bias.
6. Explain biasing.
7. Explain diode biasing.
8. What is current-mirror biasing?

9. Discuss different types of biasing techniques used in the case of BJT amplifiers.
10. What is stability of an amplifier?
11. What are stability factors?
12. Derive the current-stability factor in the cases of fixed and R_1/R_2 bias with R_E incorporated.
13. Derive voltage-stability factor in the case R_1/R_2 bias.
14. Derive current-gain stability factor in the case R_1/R_2 bias.
15. Explain thermal runaway.
16. Analyse the current-mirror circuit.
17. Explain dual-supply biasing.
18. What do you mean by self bias?
19. What are h-parameters?
20. Derive the small-signal hybrid equivalent circuit of the bipolar junction transistor.
21. Demonstrate how h-parameters can be computed form VI characteristics.
22. Explain the frequency response of BJT.
23. Derive the hybrid-π equivalent circuit of the BJT.
24. Bring-out the equivalence between the hybrid and hybrid-π parameters.
25. Compare BJT and MOSFET?
26. Give a few applications of BJT.

UNSOLVED PROBLEMS

P2.1 Compute the value of current gain in the common-base configuration, if the collector current $I_C = 1$ mA and the emitter current $I_E = 1.1$ mA.

[**Hint:** use the relation $\alpha = I_C/I_E$].

P2.2 Compute the value of current gain in the common-emitter configuration, if the collector current $I_C = 1.0$ mA and the emitter current $I_E = 1.01$ mA. Compute the value of α in this case.

[**Hint:** use the relations $\beta = I_C/I_B$, and $\alpha = \beta/1 + \beta$]

P2.3 Compute the value of current gain in the common-emitter configuration, if the collector current I_C = 1.0 mA and the emitter current I_E = 1.01 mA. Compute the values of α and γ also.

[**Hint:** use the relations $\beta = I_C/I_B$, and $\alpha = \beta/1+\beta$, and $\gamma = 1+\beta$]

P2.4 In a certain measurement on hybrid parameters, the following tabulations are obtained. Compute all the relevant hybrid parameters (I_C in mA, I_B in μA; V_{BE} and V_{CE} in V).

V_{CE}	V_{BE}	I_B
	0.00	0
	0.40	20
5	0.50	35
	0.60	55
	0.70	75
	0.00	0
	0.40	20
10	0.50	30
	0.60	40
	0.70	55

I_B	V_{CE}	I_C
	0	0
	5	1
10	10	2
	15	3
	20	4
	0	0
	5	2
20	10	4
	15	6
	20	8

P2.5 Compute the value of short-circuit current gain in the common-base configuration, if the change in collector current ΔI_C = 1 mA corresponding to a change in emitter current ΔI_E = 1.1 mA.

[**Hint:** use the relation $\alpha = \Delta I_C/\Delta I_E$]

P2.6 Compute the value of short-circuit current gain, given that at V_{CB} = 5 V, for I_E = 1.01 mA, I_C = 1.0 mA and for I_E = 2.02 mA, I_C = 2.0 mA. Specify the configuration used.

[**Hint:** use the relation $\alpha = \Delta I_C/\Delta I_E$].

P2.7 Compute the value of short-circuit current gain, given that at V_{CE} = 10 V, for I_B = 10 μA, I_C = 1.0 mA and for I_B = 15 μA, I_C = 2.0 mA. Specify the configuration used.

[**Hint:** use the relation $\beta = \Delta I_C/\Delta I_B$].

P2.8 Compute the value of reverse voltage transfer ratio, given that at I_E = 1 mA, for V_{CB} = −5 V, V_{EB} = 0.75 V, and for V_{CB} = −10 V, V_{EB} = 0.55 V. Specify the configuration used.

[**Hint:** *CBC*, $h_{rb} = \Delta V_{CB}/\Delta V_{EB}$].

P2.9 Compute the value of reverse voltage transfer ratio, given that at I_B = 15 μA, for V_{CE} = 5 V, V_{BE} = 0.55 V, and for V_{CE} = 10 V, V_{EB} = 0.7 V. Specify the configuration used.

[**Hint:** *CEC*, $h_{re} = \Delta V_{CE}/\Delta V_{BE}$].

CHAPTER 3

BIASING METHODS

3.1 INTRODUCTION

Bipolar junction transistors can be used to amplify signals when their *EB* junction is forward-biased and *CB* junction reverse biased. We thus require two separate DC power supplies for the operation of an amplifier. The common practice is to derive both these DC supply voltages from a single 230-V, 50-Hz AC power-supply using rectifiers and voltage stabilisers.

3.2 BIASING METHODS

Biasing means connecting appropriate external DC supply voltages between respective terminals of an active device for its proper operation in the desired mode. For example, if a BJT is to be operated as an amplifier, we have seen that it must have its *EB* junction forward biased and *CB* junction reverse biased. Similarly, if the BJT is to be operated as a switch, it must be biased so that it will be in its saturation or cut-off conditions. BJT biasing techniques include:

- Base (fixed) biasing
- Collector-to-base biasing
- Voltage-divider biasing
- Diode biasing.

Before we start the discussion on various biasing techniques, let us explain about operating point.

3.2.1 Operating Point

Figure 3.1 shows the biasing arrangements required to make the *EB* junction forward biased and *CB* junction reverse biased. For proper amplifier operation, the base-to-emitter voltage V_{BE} should be around 0.65 V, and the collector-to-emitter voltage V_{CE} = 5 V. These voltages are called the respective *BE* and *CE* biases.

DC condition is the condition reached by the amplifier immediately after it has been switched on. In this condition, the base-emitter voltage V_{BE} (= 0.65 V) will drive a fixed DC base current I_B.

The intersection V_{BE} with the corresponding value of I_B, defined under the steady-state condition is known as **DC operating (quiescent-*Q*)** point in the base-emitter circuit. Figures 3.2 show the operating points marked on the input characteristics.

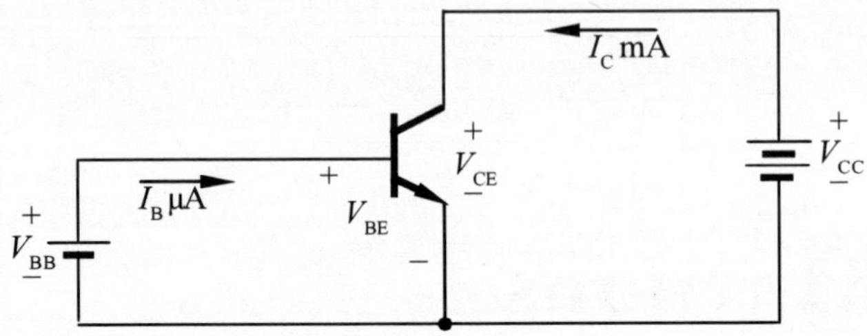

Fig. 3.1 BJT amplifier in *CEC*.

Figure 3.2 shows the input (V_{BE}-vs-I_B) characteristic of a BJT. The Q-point is fixed in the middle of its active region, as shown. The corresponding base-emitter voltage and current are, respectively, $V_{BE} = V_{BEQ}$ and $I_B = I_{BQ}$. For a typical low-power BJT, these values may be taken as 0.65 V and 15 μA, respectively. In Fig. 3.2, we observe that I_B starts linearly from $I_B = I_{Bmin}$ (0 mA) at $V_{BE} = V_{BEmin}$ (0.5 V) to $I_B = I_{Bmax}$ (30 μA) at $V_{BE} = V_{BEmax}$ (0.3 V).

Let us now apply an input signal of $v_{be} = V_{max} \sin \omega t = 0.15 \sin \omega t$ volt, as shown in Fig. 3.3. The actual input between base and emitter then becomes the sum of the DC bias voltage and AC signal voltage, i.e., $0.65 \pm 0.15 \sin \omega t$. Correspondingly, I_B varies from the minimum value of 0 μA to the maximum value of 30 μA. Mathematically, we may express this as $I_B = A \times 0.15 \sin \omega t$, where A is a constant relating I_B to V_{BE}.

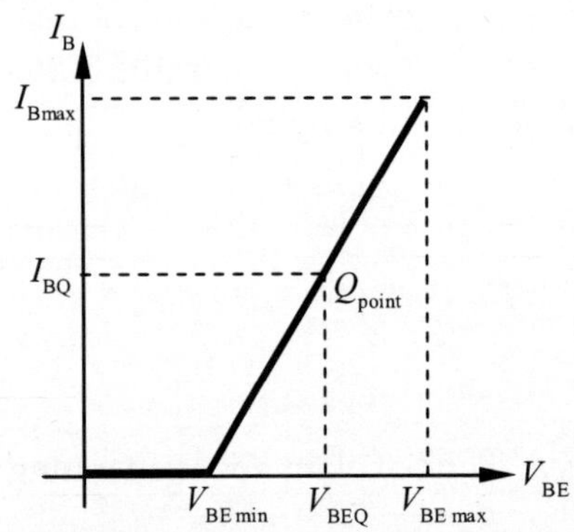

Fig. 3.2 Q point in the base-emitter region.

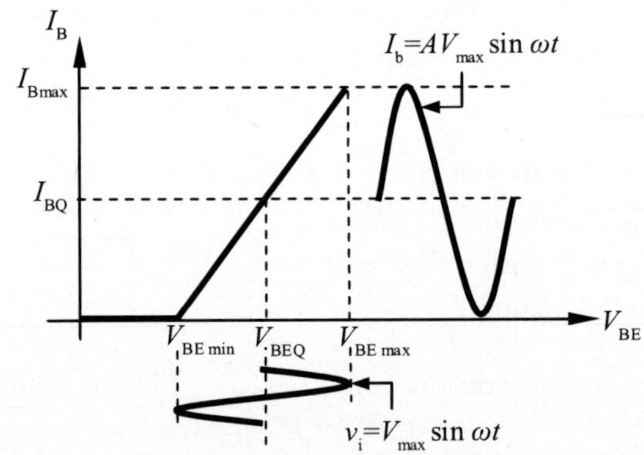

Fig. 3.3 Input waveforms.

It can be seen that, if the Q-point is in the middle of the active region, the swings in the input current and voltage are a maximum. This in turn results in a maximum swing in the collector current I_C, since $I_C = \beta I_B$, where β is the common-emitter current amplification factor, a constant for a given transistor. This is illustrated in Fig. 3.4.

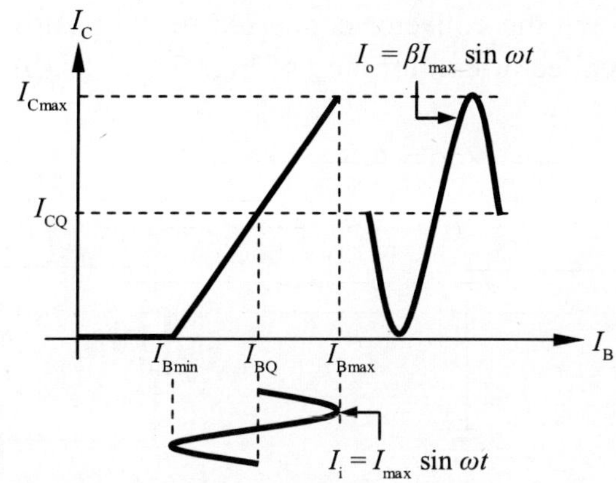

Fig. 3.4 Input/output waveforms.

Figure 3.5 shows the fixing of Q point in the output characteristics. We assume the output characteristics to be linear and equidistant from each other. The characteristics shown are for $I_B = 5$, 10, and 15 μA, respectively.

Load line is defined as *the line joining the supply voltage* (V_{CC}) *and the maximum possible value of collector current* (I_{CMAX}), as shown in Fig. 3.5. It represents the variation of collector current with respect to the collector-emitter voltage. As can be seen from Fig. 3.5, when V_{CE} is increasing, I_C is decreasing and vice versa. This means that V_{CE} and I_C are 130° out-of phase with each other.

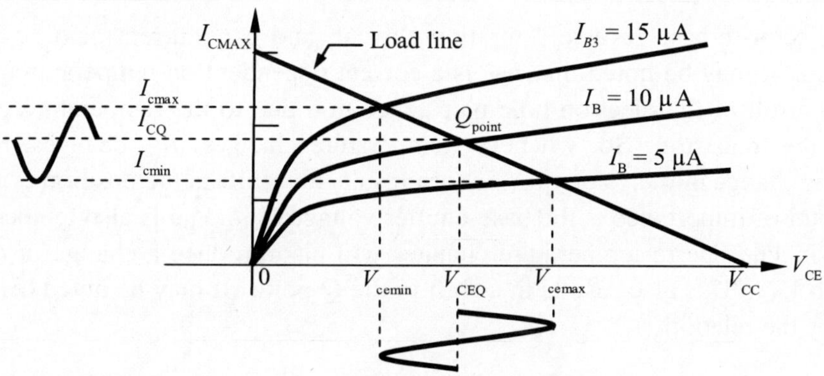

Fig. 3.5 Fixing Q-point on the output characteristics (CEC).

3.3 BIAS TECHNIQUES

3.3.1 Base (Fixed) Bias

In the base- or fixed-bias arrangement shown in Fig. 3.6(a), we use a base resistor R_B to fix the operating point in the base circuit, and a collector resistor R_C to fix the operating point in the collector circuit. Figure 3.6(b) shows a standard method of drawing the circuit shown in Fig. 3.6(a).

In this drawing, we have shown the collector connected to the positive terminal of V_{CC} and the emitter grounded. From the collector-emitter loop of Fig. 3.6(a) or (b), using Kirchhoff's voltage law that

$$V_{CC} - I_C R_C - V_{CE} = 0 \tag{3.1}$$

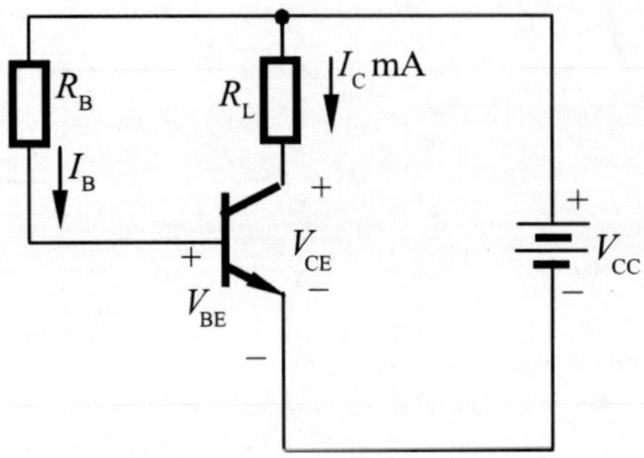

Fig. 3.6(a) Fixed bias.

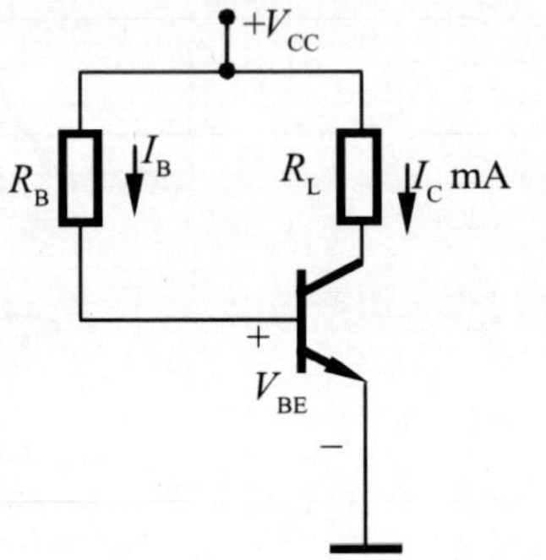

Fig. 3.6(b) Fixed bias (standard drawing).

Thermal Stability

Thermal Stability of an amplifier is the stability of its Q-points against the variations in the ambient temperature as well as the temperature developed due to current flow within the circuit. We know that the collector current of a *CE* amplifier may be written as:

$$I_C = \alpha I_E + I_{CO} \tag{3.2}$$

where I_{CO} = collector-to-base reverse saturation current, and α = current gain in the common-base configuration. It may be noted that I_{CO} is a current dependent on temperature. This current is produced as a result of the electron-hole pair generation due to the temperature present in the atmosphere and the transistor. So, whenever temperature changes, I_{CO} changes, resulting in a change in I_C. This change in the value of I_C in turn results in a shift in the position of the Q-point.

A second factor of importance is the base-emitter voltage V_{BE}. V_{BE} is also temperature dependent. A change in V_{BE} due to temperature changes will also produce a change in I_C, since it is directly related to V_{BE}. This also results in a shift in the Q-point. It may be noted here that I_C and V_{BE} are related by the relation

$$I_C = g_m V_{BE} \tag{3.3}$$

where g_m is the transconductance of the BJT.

A third factor, which is important from the stability of the Q-point, is the current-amplification factor β. β is also temperature dependent. However, the problem with β is not because of its temperature dependency. The real problem lies in the fact that for the same type of transistor, the value of β varies as large as 1 : 4. For example, we find that β of BC 107 type of transistor lies between 125 to 500. This means that the manufacturer cannot guarantee that two BC 107 transistors will have the same value of β. This problem becomes severe when we have to replace a faulty transistor with a good one. In such a case, a shift in the Q-point is sure to occur. The shift in the Q-point is usually corrected by means of variable potentiometers or resistors.

Stability Analysis using Stability Factors

In general, we may express the collector current

$$I_C = f(I_{CO}, V_{BE}, \beta, V_{CC}, \ldots) \tag{3.4}$$

A change in I_C can now be written as

$$\Delta I_C = \left[\frac{\partial I_C}{\partial I_{CO}}\right] \times \Delta I_{CO} + \left[\frac{\partial I_C}{\partial V_{BE}}\right] \times \Delta V_{BE} + \left[\frac{\partial I_C}{\partial \beta}\right] \times \Delta\beta + \left[\frac{\partial I_C}{\partial V_{CC}}\right] \times \Delta V_{CC} + \ldots \tag{3.5}$$

where the factors in the brackets, called **stability factors**, are defined as

$$\text{Current stability factor } S_I = \partial I_C / \partial I_{CO} \tag{3.6}$$

$$\text{Voltage stability factor } S_V = \partial I_C / \partial V_{BE} \tag{3.7}$$

$$\text{Current-gain stability factor } S_\beta = \partial I_C / \partial \beta \tag{3.8}$$

Current Stability Factor S_I

From the base-emitter loop using KVL, we obtain

$$V_{CC} - I_B R_B - V_{BE} = 0 \tag{3.9}$$

We also have the relation

$$I_C = \alpha I_E + I_{CO} \tag{3.10}$$

Equation (3.10) may also be written as

$$I_C = [\beta/(1+\beta)]\, I_E + I_{CO} \tag{3.11}$$

where we have used the relation

$$\alpha = \beta/(1+\beta) \tag{3.12}$$

But, we know that

$$I_E = I_C + I_B \tag{3.13}$$

Hence, substituting for I_E from Eq. (3.13) into Eq. (3.11), we get

$$I_C = [\beta/(1+\beta)]\,(I_C + I_B) + I_{CO} \tag{3.14}$$

Rearranging Eq. (3.14), we have

$$I_C = \beta I_B + (1+\beta)\, I_{CO} \tag{3.15}$$

Equation (3.15) can be rearranged to give

$$I_B = (I_C/\beta) - [\beta/(1+\beta)]\, I_{CO} \tag{3.16}$$

Substituting for I_B from Eq. (3.16) into Eq. (3.9) yields

$$V_{CC} - V_{BE} - (I_C/\beta) R_B - [\beta/(1+\beta)]\, I_{CO} R_B = 0 \tag{3.17}$$

Differentiating Eq. (3.17), assuming V_{CC}, V_{BE}, and R_B to be constants, we get

$$\partial I_C = (1+\beta)\, \partial I_{CO} \tag{3.18}$$

From Eq. (3.18), we find

$$S_I = \partial I_C/\partial I_{CO} = 1 + \beta \tag{3.19}$$

In Eq. (3.19), assume that $\beta = 100$. Then

$$S_I = \partial I_C/\partial I_{CO} = 1 + 100 = 101 \tag{3.20}$$

From Eq. (3.20), we observe that even a small change in I_{CO} (due to change in temperature) will produce a very large (in this case, 101 times) change in I_C. This in turn produces a large shift in the Q-point, and results in the instability of the amplifier.

Apart from the reason given above, another reason for instability is that, in the fixed-bias circuit, the base current I_B is directly derived from V_{CC} through resistor R_B. Since I_B is highly temperature dependent, the variations in I_B will shift the Q-point.

Yet another reason for the instability in fixed-bias circuits is that the base-emitter voltage V_{BE} is derived from V_{CC} according to the equation

$$V_{BE} = V_{CC} - I_B R_B \tag{3.21}$$

Equation (3.21) shows that since V_{BE} is also derived from I_B, it will be highly unstable, resulting in a shift in the Q-point. Due to these reasons, the fixed-bias circuit is not used for constructing BJT amplifiers.

3.3.2 Voltage-Divider-Bias

For more stability, we use a voltage-divider network consisting of resistors R_1 and R_2, as shown in Fig. 3.7 (a). The base-supply voltage can be obtained from Fig. 3.7 (a) as

$$V_{BB} = V_{CC} R_1/(R_1 + R_2) \tag{3.22}$$

and the base equivalent resistance

$$R_B = R_1 R_2/(R_1 + R_2) \tag{3.23}$$

Using these values, we draw the equivalent circuit shown in Fig. 3.7 (b).

Since V_{CC}, R_1 and R_2 are constants and temperature independent, we find that V_{BB} is also constant and temperature independent. Further, we find that I_B is not derived directly from V_{CC}, but is rather derived from the current I_R, which flows through the resistors R_2 and R_1 and is given by

$$I_R = V_{CC}/(R_1 + R_2) \tag{3.24}$$

It can be seen that I_R is also constant and temperature independent, just like V_{BB}. Now, we find from Fig. 3.3-b that base current I_B is derived from I_R. If I_B is designed to be only a very small percentage (say, 1%) of I_R, then we can see that the variations created in I_B due to fluctuations in temperature will be negligibly small. The same reasoning goes for changes in V_{BE}. So, for a practical BJT amplifier, we use the voltage-divider-bias circuit, in association with the emitter resistor. The voltage-divider bias, along with the emitter resistor is sometimes called the **self-bias circuit**.

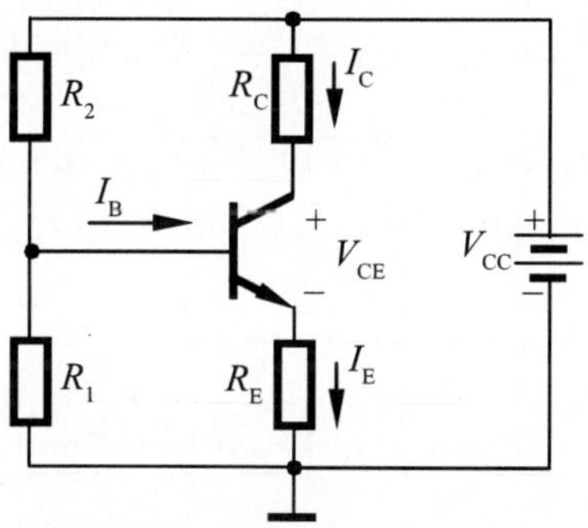

Fig. 3.7(a) Voltage divider-bias.

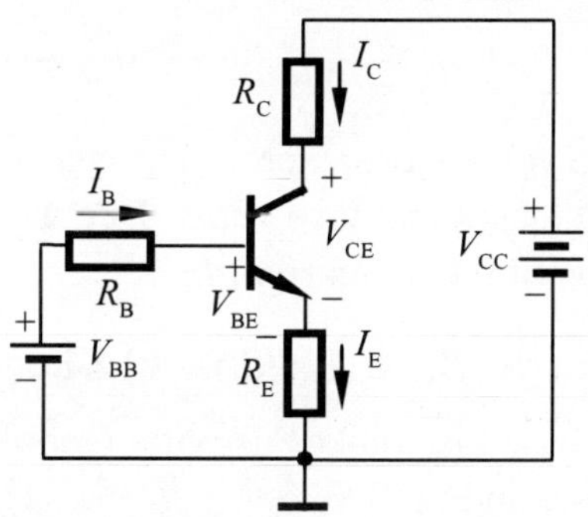

Fig. 3.7(b) Equivalent of circuit of Fig. 3.7 (a).

3.3.3 Thermal Runaway

Consider the expression for collector current of a transistor given by

$$I_C = \alpha I_E + I_{CO}$$

In this equation, the term I_{CO} is, as we have seen, temperature dependent. When the atmospheric temperature rises, I_{CO} increases, which in turn increases I_{CO}. An increase in I_C is followed by a heating of the collector-base junction of the transistor; this heating of the junction in turn produces more electron-hole pairs resulting in a further increase in I_{CO} and hence in I_C. This is a cumulative process, and we find that I_C rises to a very high value in a very short period of time, which results in a high junction temperature. This high junction temperature heats up the transistor, which ultimately results in the destruction of the transistor.

The process of a transistor getting heated up due to an increase in the (atmospheric) temperature, and hence in the collector current, which results in a shift in the Q-point and ultimately to the destruction of the transistor, is called **thermal runaway**.

Consider Fig. 3.8. Assume that I_C increases due to an increase in temperature. This in turn results in an increase in the voltage drop across R_E.

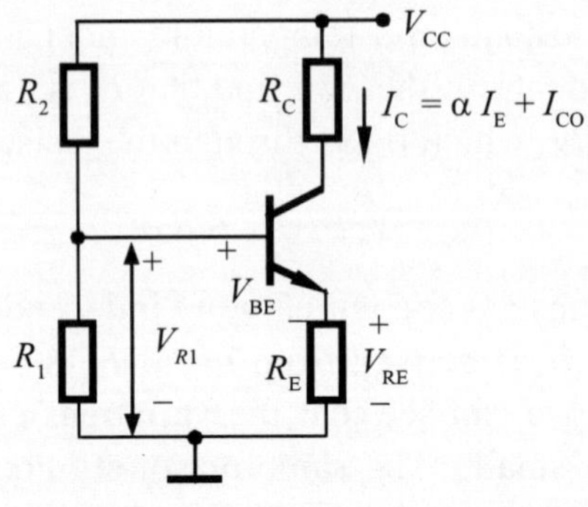

Fig. 3.8 Preventing thermal runaway.

From Fig. 3.8, we get from the base-emitter loop

$$V_{R1} = V_{RE} + V_{BE}$$

With V_{R1}, the voltage drop across R_1, is a constant (since R_1, R_2 and V_{CC} are constants) as V_{RE} increases V_{BE} decreases and vice versa. But, a decrease in V_{BE} results in a decrease in I_C, resulting in the cancellation of the increase in I_C.

3.3.4 Collector-to-Base (Feedback) Bias

The collector-to-base bias makes use of a feedback resistor *RB* between the collector and base of a transistor, as shown in Fig. 3.9.

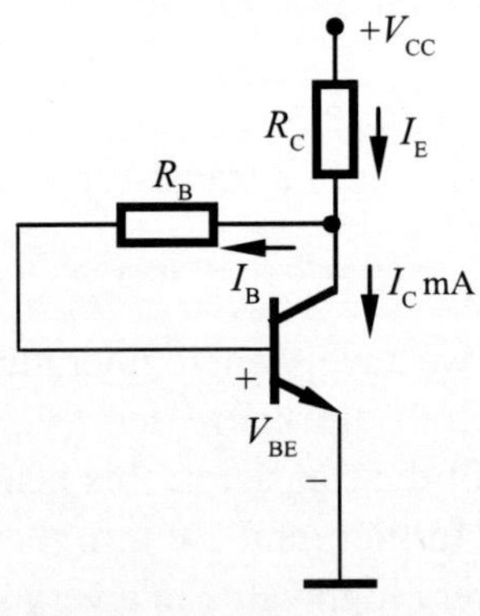

Fig. 3.9 Collector-to-base bias.

Applying KVL around the loop, we obtain the equation

$$V_{CC} - I_E R_C - I_B R_B - V_{BE} = 0 \tag{3.25}$$

Notice that the current flowing through $R_C = I_E$ and not I_C; as can be seen from Fig. 3.9, at the collector junction, emitter current I_E flowing through collector resistor R_C gets divided as base current I_B flowing through the base resistor R_B and collector current I_C flowing through the transistor.

Equation (3.25) may be rewritten as

$$V_{CC} - V_{BE} = (I_C + I_B) R_C + I_B R_B \tag{3.26}$$

Using the relation $I_C = \beta I_B$ in Eq. (3.26), we obtain the value of the base current

$$I_B = \frac{V_{CC} - V_{BE}}{(\beta + 1) R_C + R_B} \tag{3.27}$$

Stability factor analysis with Collector-to-Base Bias

We may also express Eq. (3.26) in the form

$$V_{CC} - V_{BE} = I_C R_C + I_B (R_B + R_C) \tag{3.28}$$

Now, substituting the relation

$$\frac{I_C}{\beta} - \frac{1 + \beta}{\beta} I_{CO} = I_B$$

in Eq. (3.28), we get

$$\left(\frac{I_C}{\beta} - \frac{1 + \beta}{\beta} I_{CO}\right)(R_C + R_B) + I_C R_C = V_{CC} - V_{BE} \tag{3.29}$$

Rearranging Eq. (3.29), we find

$$I_C [R_B + (1 + \beta) R_C] - [(1 + \beta) I_{CO} (R_C + R_B)] = \beta (V_{CC} - V_{BE}) \tag{3.30}$$

Differentiating both sides of Eq. (3.30) yields the currents stability factor

$$S_I = \frac{\partial I_C}{\partial I_{CO}} = \frac{[(1 + \beta)(R_C + R_B)]}{[R_B + (1 + \beta) R_C]} = \frac{(1 + \beta) R_C [1 + (R_B/R_C)]}{(1 + \beta) R_C \{1 + [R_B/(1 + \beta) R_C]\}} \cong 1 + (R_B/R_C) \tag{3.31}$$

Since R_B/R_C is small, we find the temperature stability to be good with the collector-to-base feedback bias.

SOLVED PROBLEMS

S3.1 Compute the values of I_C, I_B, and V_{CE} in the circuit shown in Fig. S3.1, given $\beta = 100$.

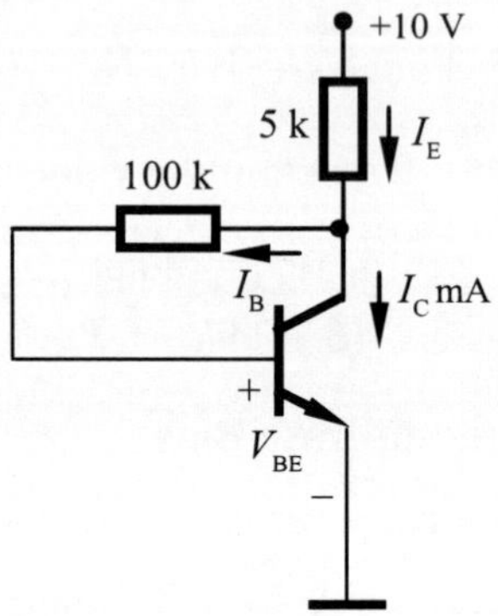

Fig. S3.1 Collector-to-base bias.

Solution

Using Eq. (3.27), we find

$$\mathbf{I_B = \frac{V_{CC} - V_{BE}}{(\beta + 1) R_C + R_B} = \frac{10 - 0.7}{101 \times 5\,k + 100\,k} = 15.37 \mu A}$$

$$\mathbf{I_C = \beta I_B = 100 \times 15.37\ \mu A = 1.537\ mA}$$

$$\mathbf{V_{CE} = V_{CC} - I_C R_C = 10 - 1.537\ mA \times 5\ k = 2.3\ V}$$

S3.2 Compute the maximum and minimum values of I_C, and V_{CE} in the circuit shown in Fig. S3.1, given that $\beta_{min} = 100$ and $\beta_{max} = 150$ and $R_C = 3.3$ k.

Solution

Using Eq. (3.27), we have found that

$$I_B = 15.37\ \mu A$$

With $\beta_{min} = 100$, we have also computed

$$\mathbf{I_{Cmin} = 1.537\ mA}$$

$$\mathbf{V_{CEmax} = V_{CC} - I_{Cmin} R_C}$$

$$\mathbf{= 10 - 1.537 mA \times 3.3 k = 4.93\ V}$$

With $\beta_{max} = 150$, we now find

$$\mathbf{I_{Cmax} = \beta_{max} I_{Bmax} = 150 \times 15.37\ \mu A = 2.31\ mA}$$

$$\mathbf{V_{CEmin} = V_{CC} - I_{Cmax} R_C = 10 - 2.31 mA \times 3.3 k = 2.377\ V}$$

S3.3 Compute the values of V_B, V_{CE}, V_C, and V_E in the circuit shown in Fig. S3.3.

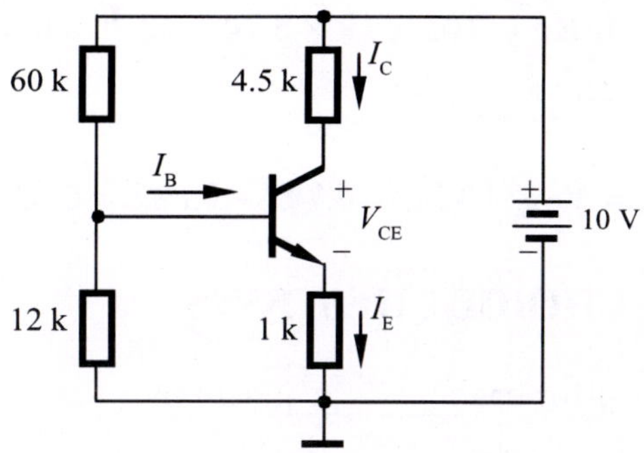

Fig. S3.2

Solution

Using Eq. (3.22), we obtain

$$\mathbf{V_{BB} = V_B = V_{CC}R_1/(R_1 + R_2) = \frac{10 \times 12k}{12k + 60k} = 1.67\ V}$$

and the base equivalent resistance using Eq. (3.23) is obtained as

$$R_B = R_1R_2/(R_1 + R_2) = \frac{12 \times 60}{12 + 60} \times k = 10\ k$$

The base current can be obtained by writing KVL around the base loop. Thus, we have

$$V_{BB} - V_{BE} = I_BR_B + I_ER_E = I_B\,(R_B + R_E) + I_CR_E$$

Since $I_C = \beta I_B$, we find

$$V_{BB} - V_{BE} = I_B\,(R_B + R_E) + \beta I_BR_E = I_B\,[R_B + (1 + \beta)\,R_E]$$

We thus get

$$I_B = \frac{V_{BB} - V_{BE}}{R_B + (1 + \beta)\,R_E} = \frac{1.67 - 0.7}{10\ k + 101 \times 1\ k} = 8.74\ \mu A$$

$$I_C = \beta I_B = 0.874\ mA$$

Now, writing KVL in the collector loop, we find

$$V_{CC} - I_CR_C - V_{CE} - I_ER_E = 0$$

Substituting given values in the above equation, and noting that $I_E = (1 + \beta)\,I_B$, we obtain

$$\mathbf{V_{CE} = 10 - (0.874 \times 4.5) - \left(101 \times 1k \times 8.74 \times 10^{-6}\right) = 4.747\ V}$$

The emitter voltage is the drop across the emitter resistor R_E and this is given by

$$\mathbf{V_{RE} = I_E R_E = 101 \times 1k \times 8.74 \times 10^{-6} = 0.883V}$$

The collector voltage is given by

$$\mathbf{V_C = V_{RE} + V_{CE} = 4.747 + 0.883 = 5.63V}$$

ONE-WORD/MULTIPLE-CHOICE QUESTIONS

M3.1 An alloy-junction transistor has ________ junctions.

M3.2 A planar-junction transistor has ________ junctions.

M3.3 The SiO_2 layer grown on top of the *P* or *N* region of a diode is called the surface ________ layer.

M3.4 For proper operation the *E-B* junction of a transistor must be ________ biased and its ________ junction must be ________ biased.

M3.5 A transistor under saturation has both the junctions ________ biased.

M3.6 The three operating regions of a transistor *VI* characteristics are ________ and ________ respectively.

M3.7 In the common-base operation, the collector current can be made zero, only if the *CB* junction is ________ biased.

M3.8 In the common-base operation, the collector current is zero when $V_{CB} = 0$. (True/False)

M3.9 When V_{CB} is increased, I_C is increased due to a phenomenon called base-width ________.

M3.10 When V_{CB} is increased, I_C is increased due to an effect called the ________ effect.

M3.11 The Early voltage is a ________ quantity.

M3.12 Collector current will flow even when $V_{CB} = 0$. (True/False)

M3.13 Collector current will not flow when $V_{CB} = 0$. (True/False)

M3.14 For a *CBC*, the input impedance is ________.

M3.15 For a *CCC*, the input impedance is ________.

M3.16 For a *CEC*, the input impedance is ________.

M3.17 For a *CBC*, the output impedance is ________.

M3.18 For a *CCC*, the output impedance is ________.

M3.19 For a *CEC*, the output impedance is ________.

M3.20 For a *CBC*, the current gain is ________.

M3.21 For a *CCC*, the current gain is ________.

M3.22 For a *CEC*, the current gain is ________.

M3.23 For a *CBC*, the voltage gain is ________.

M3.24 For a *CCC*, the voltage gain is ________.

M3.25 For a *CEC*, the voltage gain is ________.

M3.26 For a *CBC*, the power gain is ________.

M3.27 For a *CCC*, the power gain is ________.

M3.28 Current amplification factors in the *CB* and *CE* configurations are related as ________.

M3.29 Current amplification factors in the *CC* and *CE* configurations are related as ________.

M3.30 For a *CEC*, the power gain is ________.

M3.31 A field-effect transistor works on the principle of ________ field controlling drain current.

M3.32 The situation of drain current becoming zero with large negative gate voltage is called ________.

M3.33 The situation of drain current becoming just saturated is called ________.

M3.34 The drain current under saturation is called ________ drain current.

M3.35 The slope of the drain characteristic is called ________.

M3.36 The slope of the mutual characteristic is called ________.

M3.37 The slope of the amplification characteristic is called ________.

M3.38 The word *Q*-point stands for ________.

M3.39 Another name for operating point is ________-point.

M3.40 Operating point means ________ state condition.

M3.41 The line joining the maximum values of current and voltage is called the ________ line.

M3.42 The biasing arrangement that makes use of a fixed resistor between the collector and base of a transistor is called ________ bias.

M3.43 Collector current is given by the expression $I_C = \alpha I_E +$ ________.

M3.44 I_C, V_{BE} and g_m are related by the expression ________.

M3.45 Transconductance of a BJT is defined as ________.

M3.46 Transconductance of a BJT may be approximately expressed in terms of V_T as ________.

M3.47 Current stability factor is defined as ________.

M3.48 Voltage stability factor is defined as ________.

M3.49 Current-gain stability factor is defined as ________.

M3.50 Current stability factor for the case of a voltage-divider biasing is given by ________.

M3.51 Voltage stability factor for the case of a voltage-divider biasing is given approximately by ________.

M3.52 The variation of collector current with temperature leading to the destruction of the transistor is known as ________.

M3.53 The diode biasing used in integrated circuits is known as ________.

M3.54 The biasing obtained by using a source resistance in JFETs is known as ________.

M3.55 The biasing obtained by using a voltage-divider circuit in JFETs is known ________.

M3.56 The hybrid-π equivalent of h_{ie} is ________.

M3.57 The hybrid-π equivalent of h_{fe} is ________.

M3.58 The hybrid-π equivalent of h_{re} is ________.

M3.59 The hybrid-π equivalent of h_{oe} is ________.

M3.60 The emitter resistor is used in a transistor for temperature ________.

M3.61 The voltage-divider bias is used for fixing the ________.

M3.62 The collector resistance is used for fixing the Q-point in the ________.

M3.63 Coupling capacitor is used for ________ and ________.

M3.64 Thermal voltage V_T may be computed by using the expression ________.

REVIEW QUESTIONS

1. Explain how a bipolar junction transistor can act as an amplifier.
2. What do you mean by operating point?
3. What is load line?
4. What is meant by biasing of amplifier? Why do you require it?
5. Explain fixed bias.
6. Explain biasing.
7. Explain diode biasing.
8. What is current-mirror biasing?
9. Discuss different types of biasing techniques used in the case of BJT amplifiers.
10. What is stability of an amplifier?
11. What are stability factors?
12. Derive the current-stability factor in the cases of fixed and R_1/R_2 bias with R_E incorporated.
13. Derive voltage-stability factor in the case R_1/R_2 bias.
14. Derive current-gain stability factor in the case R_1/R_2 bias.
15. Explain thermal runaway.
16. Analyse the current-mirror circuit.
17. Explain dual-supply biasing.
18. Explain the methods for biasing JFET's and MOSFET's.
19. What do you mean by self bias?
20. What are the different biasing methods for MOSFET's.
21. What are h-parameters?
22. Derive the small-signal hybrid equivalent circuit of the bipolar junction transistor.
23. Demonstrate how h-parameters can be computed form VI characteristics.

24. Explain the frequency response of BJT.

25. Derive the hybrid-π equivalent circuit of the BJT.

26. Bring-out the equivalence between the hybrid and hybrid-π parameters.

27. Compare BJT and MOSFET?

28. Give a few applications of BJT.

UNSOLVED PROBLEMS

P3.1 Compute the value of current stability factor in the case of a fixed-bias RC-coupled amplifier with $\beta = 100$, $R_C = 1$ kΩ, $R_B = 100$ kΩ and $R_E = 1$ kΩ.

P3.2 Compute the value of current stability factor in the case of an RC-coupled amplifier with voltage-divider bias having $\beta = 100$, $R_C = 1$ kΩ, $R_B = 10$ kΩ and $R_E = 1$ kΩ.

P3.3 Compute the value of voltage stability factor in the case of an RC-coupled amplifier with voltage-divider bias having $\beta = 100$, $R_C = 1$ kΩ, $R_B = 10$ kΩ and $R_E = 1$ kΩ.

P3.4 Compute the value of current-gain stability factor in the case of an RC-coupled amplifier with voltage-divider bias having $R_C = 1$ kΩ, $R_B = 10$ kΩ and $R_E = 1$ kΩ. In this case, the current gain values are $\beta_1 = 100$ for $I_{C1} = 1$ mA and $\beta_2 = 105$ for $I_{C1} = 2$ mA.

CHAPTER 4 OTHER DEVICES

4.1 SILICON-CONTROLLED RECTIFIER

4.1.1 Introduction

Bipolar junction transistors and field-effect transistors are linear devices. They can be used as amplifiers if they are operated in the linear region of their *VI* characteristics. However, if they are operated in the saturation and cut-off regions of the characteristics, they can be used as switches. However, the power-handling capacity is usually small.

We have a second class of devices that will function only as ON-OFF switches. As they are switching devices, their power-handling capacity could be made very large. Semiconductor devices called thyristors belong to this class. The name **thyristor** is derived from the words **thyr**atron and trans**istor**. Thyristors are capable of handling large currents and voltages and hence are power devices.

4.1.2 Four-Layer Shockley Diode

The first member of the thyristor family of devices is the *PNPN* (pinpin) four-layer diode, also known as the Shockley diode (named after its inventor). The structure of the Shockley diode is shown in Fig. 4.1. The Shockley diode, as can be seen from the diagram, consists of a four-layer $P_1N_1P_2N_2$ structure. The diode has an anode (A) and a cathode (K). It can be seen that there are three junctions for the diode, which are indicated as J_1, J_2, and J_3, respectively in Fig. 4.1.

In Fig. 4.1, we find that the anode is biased positively with respect to the cathode. This makes junctions J_1 and J_3 forward-biased, and J_2 reverse-biased. When the bias is low, even though J_1 and J_3 are forward-biased, since J_2 is reverse-biased, the current I_A through the diode is the reverse-saturation current of J_2. This is very low, on the order of microamperes only. However, when we raise the applied bias to a high value, we find that avalanche breakdown of the junction occurs, and I_A becomes quite large.

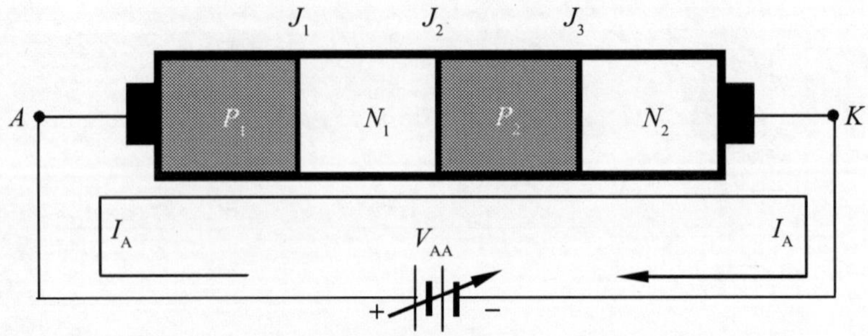

Fig. 4.1 Four-layer diode under biased condition.

4.1.3 Two-Transistor Analogy of Four-Layer Diodes

Shockley diodes can be assumed to be made up of a *PNP* transistor connected to an *NPN* transistor, as shown in Fig. 4.2. In the figure, $P_1N_1P_2$ forms the *PNP* transistor and the $N_1P_2N_2$ forms the *NPN* transistor.

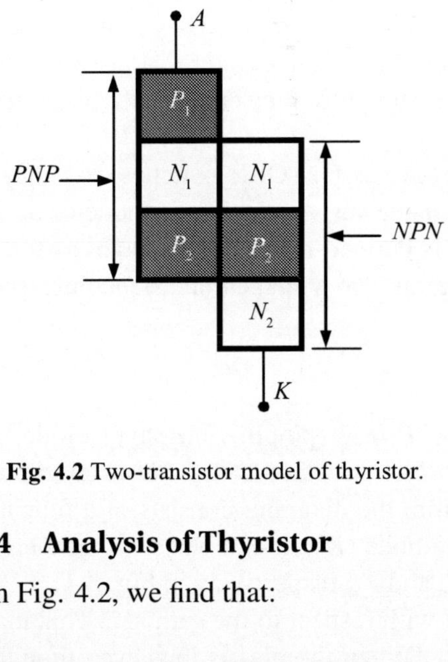

Fig. 4.2 Two-transistor model of thyristor.

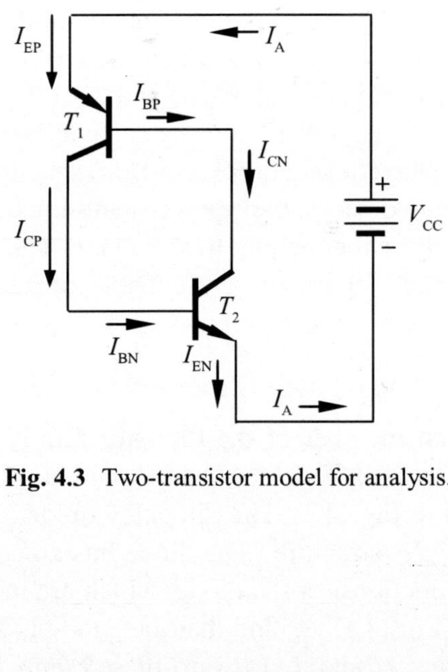

Fig. 4.3 Two-transistor model for analysis.

4.1.4 Analysis of Thyristor

From Fig. 4.2, we find that:

$$I_A = I_{EP} = I_{EN} \tag{4.1}$$

where I_{EP} = emitter current of the *PNP* transistor, and I_{EN} = emitter current of the *NPN* transistor. For a BJT, we have the relation:

$$I_C = \alpha I_E + I_{CO} \tag{4.2}$$

where I_C collector current, α = current gain in the common-base configuration, and I_{CO} = reverse saturation current. Using this expression for the *PNP* and *NPN* transistors, we have the respective collector currents given by:

$$I_{CP} = \alpha_P I_{EP} + I_{COP} \tag{4.3}$$

$$I_{CN} = \alpha_N I_{EN} + I_{CON} \tag{4.4}$$

where the subscript *P* stands for the *PNP* variety and the subscript *N* stands for the *NPN* variety. Now, substituting for I_A from Eq. (4.1) into Eq. (4.3), we have

$$I_{CP} = \alpha_P I_A + I_{COP} \tag{4.5}$$

Now, from Fig. (4.4), we get

$$I_{EP} = I_{CP} + I_{BP} \tag{4.6}$$

Substituting for I_{CP} from Eq. (4.5) into Eq. (4.6), we get

$$I_{EP} = \alpha_P I_A + I_{COP} + I_{BP} \tag{4.7}$$

Substituting for I_{EP} in Eq. (4.7), we have

$$I_A = \alpha_P I_A + I_{COP} + I_{BP} \tag{4.8}$$

Rearranging Eq. (4.8) yields

$$I_A (1 - \alpha_P) = I_{COP} + I_{BP} \tag{4.9}$$

Now, again referring back to Fig. 4.3, we find that

$$I_{BP} = I_{CN} \tag{4.10}$$

Therefore, from Eq. (4.10), we get that

$$I_A (1 - \alpha_P) = I_{COP} + I_{CN} \tag{4.11}$$

Substituting for I_{CN} from Eq. (4.4) into Eq. (4.11), we find that

$$I_A (1 - \alpha_P) = I_{COP} + \alpha_N I_{EN} + I_{CON} \tag{4.12}$$

From Eq. (4.12), by substituting for I_{EN}, we get

$$I_A (1 - \alpha_P) = I_{COP} + \alpha_N I_A + I_{CON} \tag{4.13}$$

Rearranging Eq. (4.13), we find that

$$I_A (1 - \alpha_P - \alpha_N) = I_{COP} + I_{CON} = I_{CO} \tag{4.14}$$

where I_{COP} and I_{CON} are the "half"-reverse saturation currents of "half"-transistors T_1 and T_2, respectively. Sum of these two half-currents yield the reverse saturation current I_{CO} of the Shockley diode. Now, finally, rearranging Eq. (4.14) yields

$$\mathbf{I_A = \frac{I_{CO}}{1 - (\alpha_P + \alpha_N)}} \tag{4.15}$$

Equation (4.15) gives the expression for the anode current of a four-layer diode.

Consider the situation under low anode supply voltages. In this condition, the individual current gains α_P and α_N are small for silicon devices, such that

$$\alpha_P + \alpha_N << 1 \tag{4.16}$$

Under this condition, we have Eq. (4.15) modified as:

$$I_A = I_{CO} \tag{4.17}$$

This is the **forward-blocking** or **OFF** state, which we had mentioned in our earlier discussions.

Now, let the supply voltage be increased such that

$$\alpha_P + \alpha_N \to 1 \tag{4.18}$$

Then from Eq. (4.15), we get

$$I_A = \frac{I_{CO}}{1 - 1} = \infty \tag{4.19}$$

Equation (4.19) states that when $(\alpha_P + \alpha_N)$ approaches unity, the current becomes ideally **infinity**, and practically very high. This is called the **forward-breakdown** or **ON** state.

That the value of $(\alpha_P + \alpha_N)$ is small for low supply voltages, and approaching unity at high voltages is applicable only to silicon four-layer devices. For germanium devices $(\alpha_P + \alpha_N)$ approaches unity even when the voltage is low; therefore practically germanium four-layer devices have not been constructed so far.

4.1.5 VI Characteristics of the Shockley Diode

The anode current-anode voltage characteristic of the Shockley diode is shown in Fig. 4.4. At low values of forward anode-cathode voltage V_{AKF}, anode current is I_{CO} (represented by the region OA in Fig. 4.4). As V_{AKF} exceeds the breakdown value V_{BR} (point A in Fig. 4.4), we find that I_{AF} suddenly rises to a high value called the **latching current I_L**. At the same time, V_{AKF} falls to a low value of about 0.7 V. This transition of I_A from the low value of I_{CO} to the high value of I_L occurs very fast (shown by region AB in Fig. 4.4). The low value of the anode voltage is due to the fact that after breakdown all the three junctions are found to be forward-biased. However, we find that J_1 and J_3 are forward-biased in one direction and J_2 is forward-biased in the opposite direction.

After the diode has reached the breakdown condition, if we increase V_{AKF} further, we find that the characteristic follows that of the ordinary semiconductor diode. This is shown by the region *BC* in Fig. 4.4.

To turn the diode OFF, we have to reduce I_{AF} to a value less than I_L. It may be noted here that the device will not be turned off, if we reduce the current to just I_L. This is simply because a device that is turned on at a given current will not be turned off at the same value, as it is still ON at that current. The turn-off current value is called the **holding current I_H**.

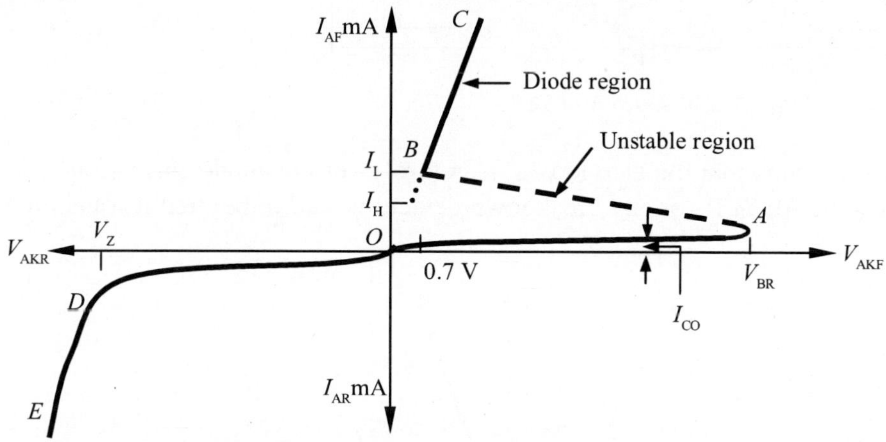

Fig. 4.4 V_{AK}-I_{AK} characteristic of Shockley diode.

The Shockley diode behaves just like an ordinary *PN* diode in the reverse-bias condition, with a comparatively larger reverse (avalanche) breakdown voltage as shown by *DE* in Fig. 4.4.

4.1.6 Silicon-Controlled Rectifier (SCR)

The most important member of the thyristor family is the silicon-controlled rectifier (SCR). SCR is a modification of Shockley diode with a gate terminal added to the P_2 region, as show in Fig. 4.5. Figure 4.6 shows the symbol of SCR.

As shown in Fig. 4.5, the gate is biased positively with respect to the cathode. A gate current I_G then starts flowing between the gate and the cathode. This current introduces holes from the P_2 region into junction J_3 and N_2 region, which results in the reduction of the overall resistance of the P_2N_2 region. The reduction in the resistance of the P_2N_2 region in turn results in a reduced breakdown voltage V_{BR1} as indicated in Fig. 4.7. We find that, as we increase the gate current I_G, the resistance of the P_2N_2 region, and hence the value of the breakdown voltage V_{BR} gets decreased correspondingly. In Fig. 4.7, we represent these as V_{BR1}, V_{BR2}, and, V_{BR3}, corresponding to gate currents I_{G1}, I_{G2}, and I_{G3}. We find that the curves generally follow the same path after breakdown, indicating that I_L and I_H remain the same for a given device. The curves also tell us that the anode current can be controlled by controlling the gate current.

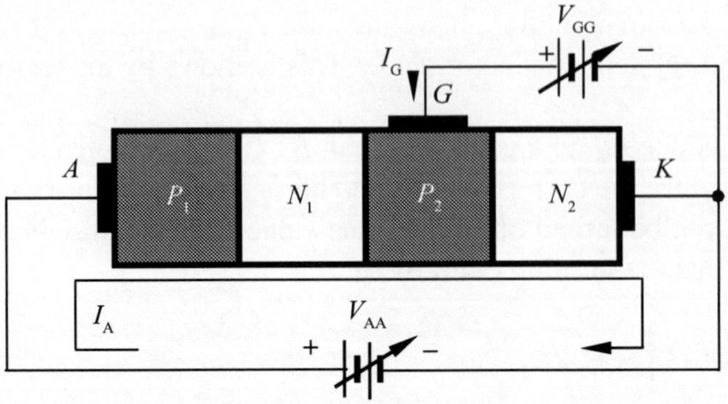

Fig. 4.5 Basic structure of SCR.

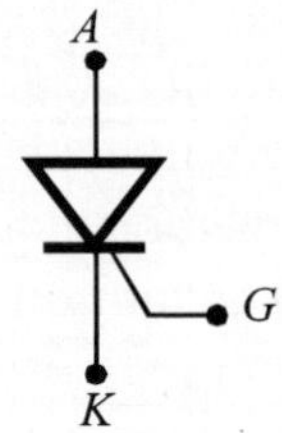

Fig. 4.6 Symbol of SCR.

However, it is found that the gate looses its control over the anode current once the device has been broken down. An SCR (or any breakdown device) is said to be fired if it is in the breakdown condition.

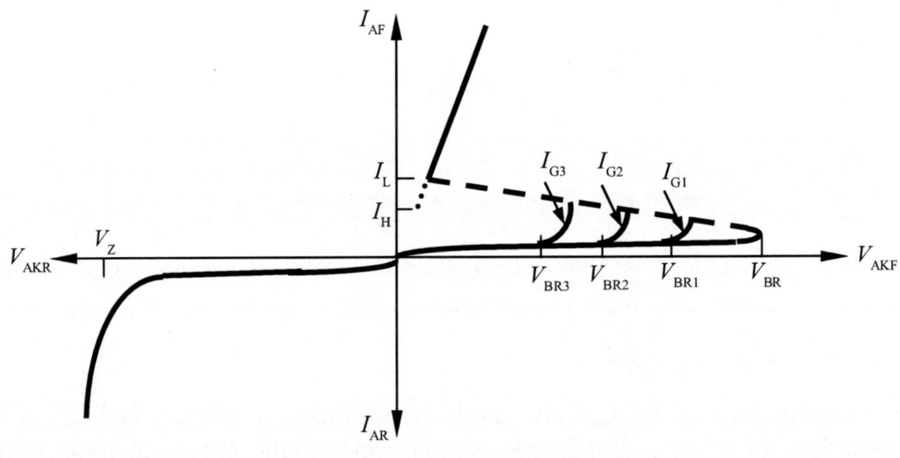

Fig. 4.7 $V_A I_A$ characteristic of SCR.

4.1.7 Triggering Methods of SCR

There are several methods by which an SCR can be fired. We have already considered the case of the DC triggering, where we have used a variable DC voltage between the gate and the cathode for firing the SCR. Below we discuss a few more types of firing of the SCR.

Phase-shift method

In the phase-shift method shown in Fig. 4.8, we use a series *RC* network to supply the current required to fire the SCR. The *RC* network is connected across the SCR, as shown. It can be also be seen that the gate of the SCR is connected the *RC* network through a diode *D*.

The working of the circuit may be explained as given below. The SCR is driven from a 230-volt AC supply. During the positive half-cycles, the capacitor C gets charged from the AC supply. When the capacitor voltage is equal to the breakdown voltage of the SCR, it fires. Heavy anode current now starts flowing through the SCR and the load connected to it. At this time, we find that the capacitor gets discharged through the SCR.

During the negative half-cycles, the anode of the SCR becomes negative with respect to its cathode. This makes the SCR to stop its conduction. This is because the negative voltage on the anode fails to attract electrons from the cathode. As a result, during the negative half-cycles, the gate regains its control over the flow of the anode current. The diode D is used to prevent the flow of reverse current through the gate of the SCR during negative half-cycles.

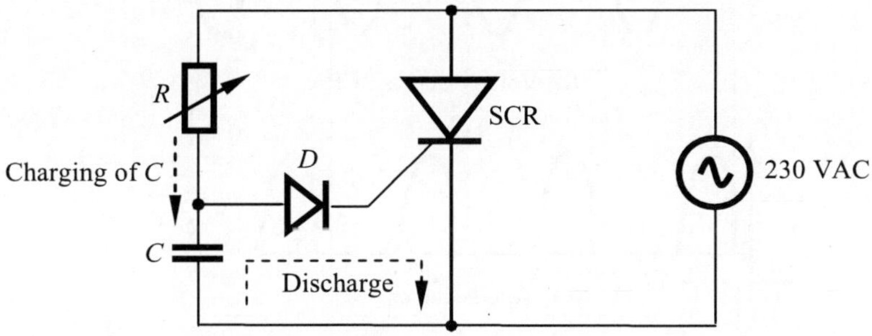

Fig. 4.8 Phase-shift method of triggering SCR.

Thus we conclude that during the positive half-cycles, the SCR gets fired and during the negative half-cycles, the SCR remains OFF. The firing angle α is given by the expression:

$$\alpha = \tan^{-1} \frac{1}{\omega CR} \tag{4.20}$$

where $\omega = 2\pi \times 50$ rad/s and α = phase angle of the applied AC voltage at which the SCR gets fired as shown in Fig. 4.9 (c).

Figure 4.9 (a) shows the applied AC input wave, Fig. 4.9 (b) shows the anode voltage, and Fig. 4.9 (c) shows the anode current waveforms. At firing point, the anode current suddenly rises and follows the applied AC wave as shown. At the same time, we find that the anode voltage drops to 0.7 V, the ON-voltage across the SCR.

From Eq. (4.20), we find that the phase angle can be varied by varying R. This varies the anode current, which means that the anode current can be controlled by varying R. This finds application in speed-control of DC motors, and AC single-phase motors.

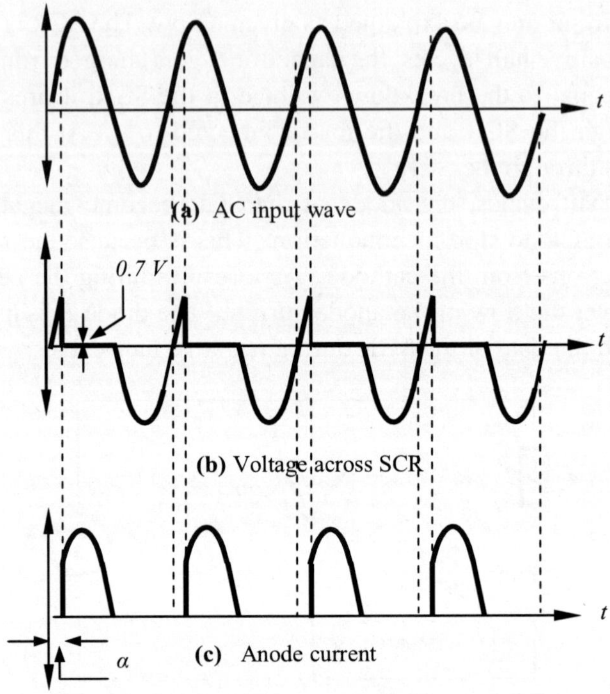

Fig. 4.9 Waveforms associated with of SCR.

4.2 UNIJUNCTION TRANSISTOR (UJT)

The structure of a typical unijunction transistor (UJT) is shown in Fig. 4.10. UJT consists of a high-resitivity N channel into which a P^{++} region is diffused. The surface is then passivated with a layer of silicon dioxide. Then windows cut into this layer, aluminium metallization done, and the emitter (E), base-1 (B_1) and base-2 (B_2) terminals are taken out, as shown.

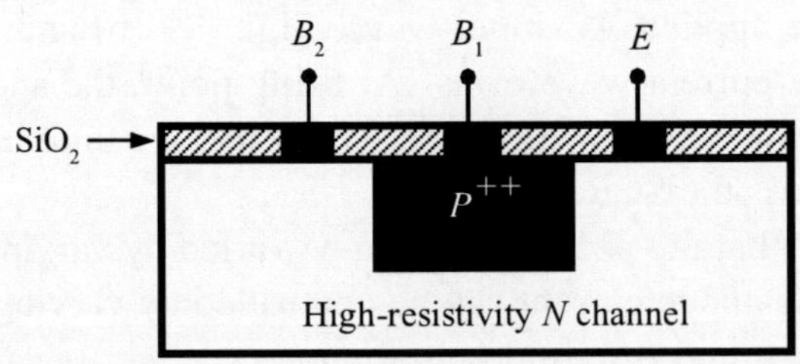

Fig. 4.10 Structure of UJT.

4.2.1 Working Principles of UJT

Figure 4.11 shows the equivalent circuit of the UJT. It consists of an emitter-base1 diode connected to a pair of resistors R_{B1} and R_{B2}, respectively. These are the actual physical resistances of the

N channel between E and the respective base terminals. The ratio of these resistances are usually expressed in the form

$$\eta = \frac{R_{B1}}{R_{B1} + R_{B2}} \tag{4.21}$$

and is called as the **intrinsic stand-off ratio**.

Figure 4.11 also shows the respective bias voltages also. Between the bases, V_{BB} is applied and a base current I_B flows between them as shown. This current develops a voltage V_{RB1} across resistance R_{B1} given by

$$V_{RB1} = \frac{V_{BB} \times R_{B1}}{R_{B1} + R_{B2}} = \eta V_{BB} \tag{4.22}$$

In Fig. 4.11, we have shown the emitter being applied with a supply voltage V_{EE}. Let us initially keep the voltage across E and B_1, $V_{EB1} \approx V_{EE} = 0$. When $V_{EB1} < V_{RB1}$, the diode D is reverse-biased, and a reverse-saturation current I_{ER} flows in the direction shown. This continues till $V_{EB1} = V_D + V_{RB1}$, where V_D = voltage drop across the diode.

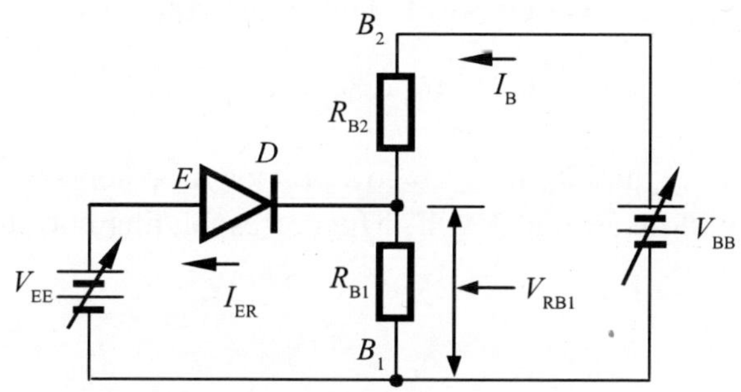

Fig. 4.11 UJT circuit with $V_{EB1} < V_{RB1}$.

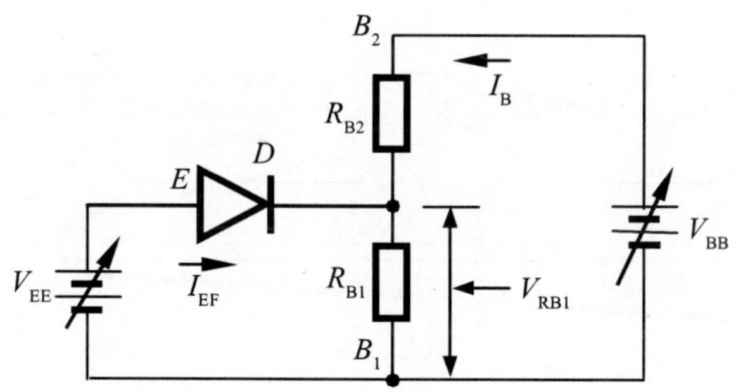

Fig. 4.12 UJT circuit with $V_{EB1} = V_{RB1}$.

When $V_{EB1} = V_{RB1}$ (see Fig. 4.12) we find that the diode becomes just forward-biased and starts to conduct in the forward direction. The emitter current I_{EF} now starts flowing from the emitter terminal through the P^{++} region and the N channel to the B_1 terminal. The action of this forward current in the channel may now be explained with the help of Fig. 4.13, which shows the characteristic of the UJT.

As V_{EB1} (in effect, V_{EE}) is increased to $\eta V_{BB} + V_D = V_P$, we find that the diode gets forward-biased and the current becomes a forward current I_{EF}, as shown. This is marked by point B in Fig. 4.12. At this point, called the **peak point**, the current I_P is on the order of μA.

Once the forward voltage across the diode D exceeds V_P, holes are injected from the P^{++} emitter region into the high-resistivity N channel. The presence of these holes in the N channel reduces its resistance between the E and B_1 regions. This reduction in the channel resistance induces the emitter to inject more holes into the channel, which further reduces the channel resistance, and induces still more holes into the channel. We find that this is a cumulative effect as a result of which the emitter current gets suddenly increased to a high value and the channel resistance gets decreased to a low value (the low channel resistance is indicated by a low voltage drop across the device). This effect with which the channel resistance gets varied is called **conductivity modulation**.

The decrease in voltage and increase in current is indicated by the region BC in Fig. 4.13. Since the slope of this region is negative, we call this a negative-resistance region. The point C reached after the UJT has fired is called the **valley point**. The valley voltage V_V can be written as

$$V_V = V_D + V_{RBI\,(low)} \tag{4.23}$$

where V_D = diode drop (0.6 V), and $V_{RBI\,(low)}$ = low voltage drop across the EB_1 region of the UJT after firing. Typically V_V will be around 1 volt. The corresponding current I_V will be on the order of a few milliamperes.

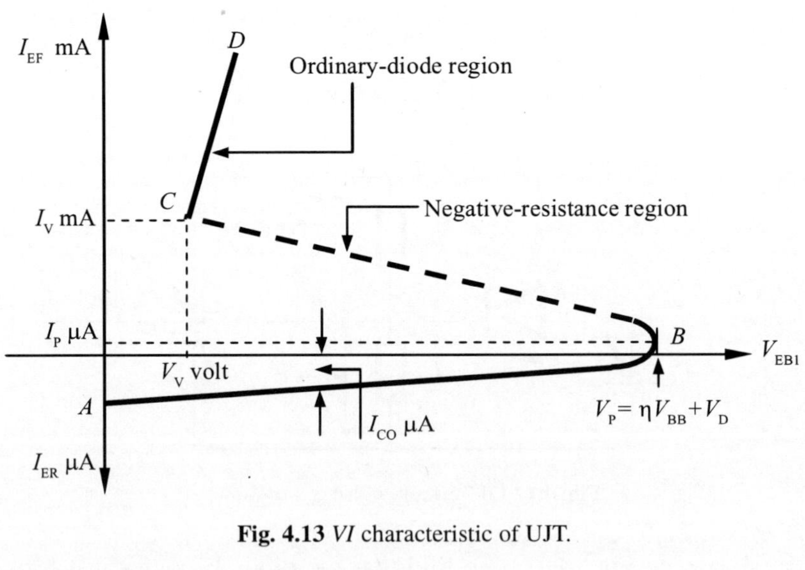

Fig. 4.13 *VI* characteristic of UJT.

After the UJT has settled in the valley point, if we increase V_{EB1}, we find that I_E increases as if the UJT were an ordinary diode. This is represented by the region CD in Fig. 4.13.

4.2.2 UJT Relaxation Oscillator

A UJT relaxation oscillator consists of a capacitor C and a resistor R connected to the emitter of the UJT, as shown in Fig. 4.14. Base B_2 of the UJT is connected to the power-supply positive through resistor R_2, and base B_1 is connected to the ground through resistor R_1. Usually R_1 and R_2 are low-value resistors, typical value being 47 ohms each. These are used to limit the current through the UJT and protect it. We can take three outputs from the UJT relaxation oscillator. These are V_{o1}, V_{o2}, and V_{o3}, respectively, as shown in Fig. 4.14. The working of the oscillator can be explained as follows:

Initially, the UJT is in the OFF-state, and the capacitor C is uncharged. C now gets charged through R towards V_{BB}, as shown. When the voltage across the capacitor

$$v_C = V_P = \eta V_{BB} + V_D \tag{4.24}$$

UJT gets fired and C gets discharged through the now ON UJT. C discharges to the valley voltage V_V, at which the UJT turns-off and C stops discharging. Once the UJT stops conducting and C discharges to V_V, we find that C again starts conducting, and the whole operation repeats. It can be seen that the charging and the discharging of the capacitor produces across C a non-linear or sawtooth waveform, as shown in Fig. 4.15.

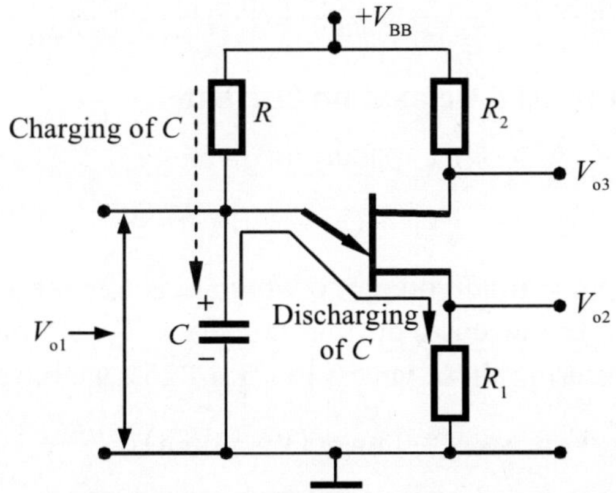

Fig. 4.14 UJT relaxation oscillator.

The waveform across C is shown in Fig. 4.15 (a). This figure shows C charging towards V_{BB} in time T_1, and discharging towards V_V in time T_2, producing a sawtooth waveform, with a period $T = T_1 + T_2$. Usually, the UJT discharges with a very low ON-resistance, so that $T_2 \ll T_1$, and hence can be neglected. Thus the period of oscillation is determined mainly by $T = T_1$. For producing oscillations, UJT operates between the peak point and the valley point. Since they are the stable operating points of the UJT, we say that the UJT operates in its relaxed modes of operation. So, we call the UJT oscillator as a *relaxation oscillator*.

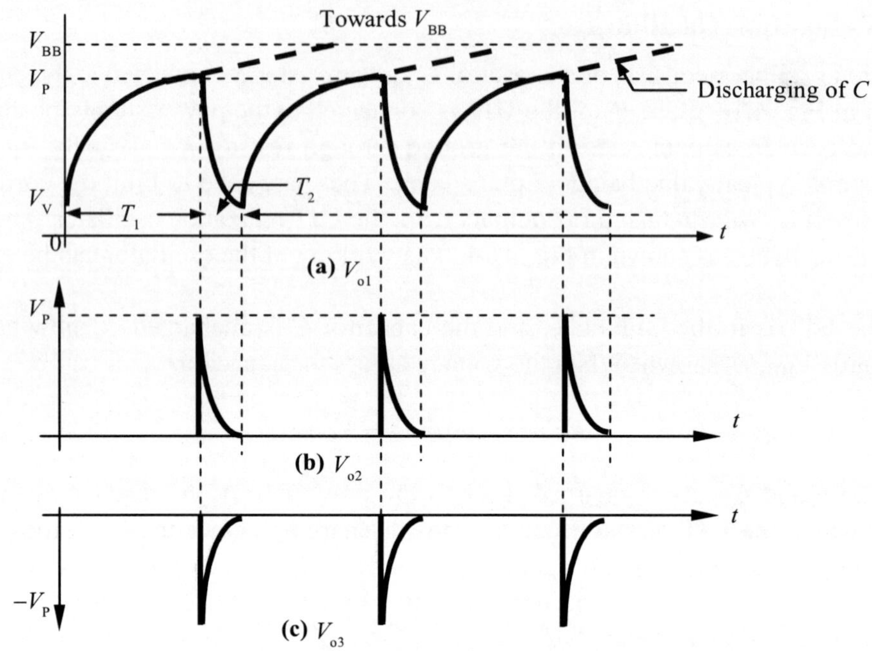

Fig. 4.15 Waveforms of UJT relaxation oscillator.

Frequency of Oscillation of the UJT Relaxation Oscillator

The equation governing the charging of a capacitor is given by

$$v_C = V_F + (V_I - V_F)\,e^{-t/RC} \tag{4.25}$$

where v_C = voltage across C, V_F = final voltage to which C is to charge to, and V_I = initial voltage from which C starts charging. In our case, during the period T_1, we have $v_C = V_P = \eta V_{BB} + V_D$, $V_F = V_{BB}$, and $V_I = V_V$. Substituting these values in Eq. (4.25), we have

$$\eta V_{BB} + V_D = V_{BB} + (V_V - V_{BB})\,e^{-t/RC} \tag{4.26}$$

Rearranging Eq. (4.26), we get the time period of oscillation as

$$t = RC\ \ln \frac{V_V - V_{BB}}{V_{BB}\,(1 - \eta) + V_D} \tag{4.27}$$

Since V_V and V_D are negligible compared to V_{BB}, Eq. (4.27) gets reduced to

$$t = T = RC \ln \frac{1}{1 - \eta} \tag{4.28}$$

Therefore, the frequency of oscillation

$$f_o = \frac{1}{RC \ln (1/1 - \eta\,)} \tag{4.29}$$

4.2.3 Applications of the UJT

- UJTs are used to produce sawtooth voltages.
- They are also used to produce sharp pulses for triggering SCRs.

4.3 JUNCTION FIELD-EFFECT TRANSISTOR (FET)

4.3.1 Structure of the JFET

Figure 4.16 shows the structure of an N-channel JFET. As shown in the figure, an N-channel JFET consists of an N region of silicon into which two P^{++} regions are diffused. The N region is called a channel because it permits current flow as in a conducting channel. The ends of the channel are metallized and external leads are taken out to form the source (S) and drain (D) terminals of the FET. The two P^{++} regions are usually shorted together to form the gate (G) terminal. We notice that immediately after the P^{++} regions are diffused into the N channel, depletion layers are formed about the P^{++} regions, as shown in Fig. 4.16.

The practical construction of an N-channel JFET is shown in Fig. 4.17. The N channel is grown as an epitaxial layer on a P^{++} substrate. The second P^{++} region is then diffused into this epitaxial layer. An SiO_2 surface-passivating layer is formed on the top, and windows are cut into this layer for aluminium metallization. The source, gate, and drain terminals are then taken out from these metallizations. It may be noted in this connection that there are two gate connections, as shown. These are tied together to form into a single gate terminal.

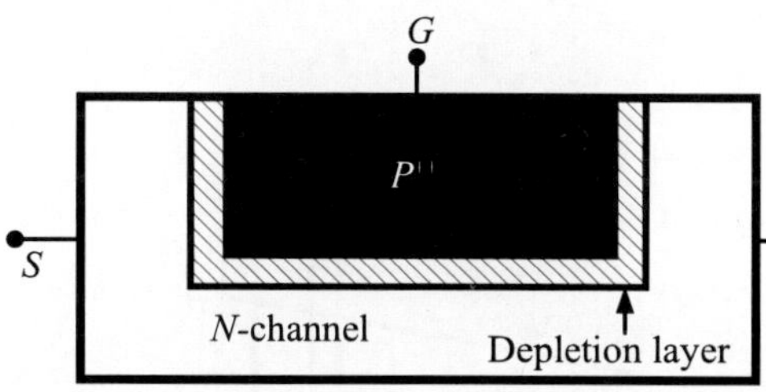

Fig. 4.16 Structure of JFET.

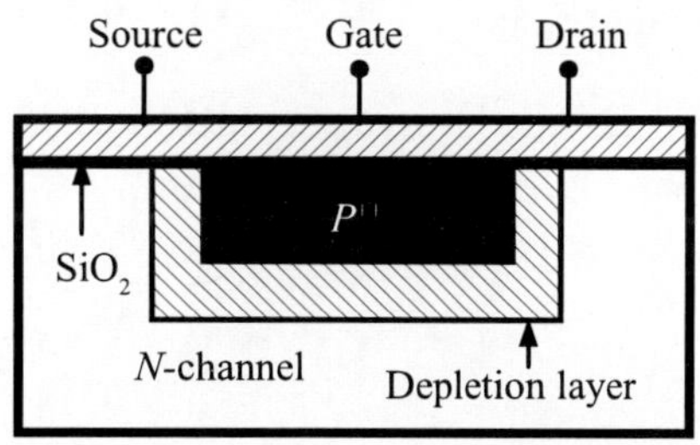

Fig. 4.17 Practical structure of JFET.

4.3.2 Working Principles of the JFET

Figure 4.18 shows the drain biased positively with respect to the source. In this condition, electrons from the source get attracted by the positive bias on the drain, which makes drain current to start flowing from the drain to the source. We also notice that the *depletion region is wider near the drain region than the source region.* This is due to the following reasons.

As stated above, in the JFET, drain current I_D flows from the drain to the source through the N channel, as if it were a conductor of finite resistance. In fact, the channel may be represented as a series connection of several resistors, each of value R ohms.

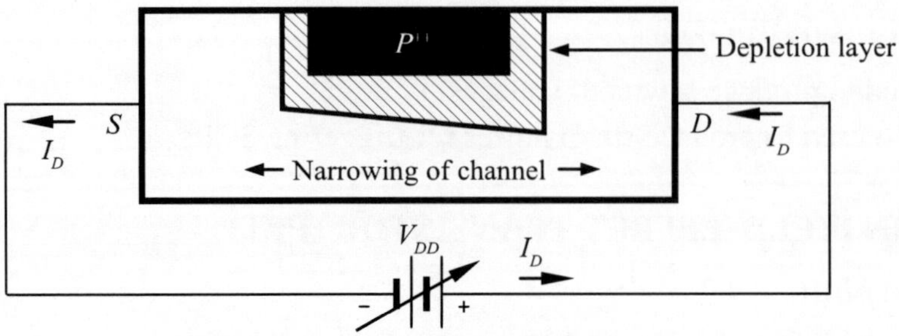

Fig. 4.18 Working of a JFET.

When the drain current I_D starts flowing through these resistors, there will be voltage drops across these resistors. These drops make the gate biased negatively with respect to the drain, and positively with respect to the source. Thus, due to the drain-current flow in the channel, we have forward bias between the source and the gate ($+V_{GS}$) and reverse bias between the gate and the drain ($-V_{DG}$). Due to the forward bias between the source and the gate, the depletion layer in this region is thin. However, due to the reverse bias, there exists a larger depletion layer between the gate and the drain, as shown in Fig. 4.18.

Now, let the supply voltage V_{DD} be increased in steps of, say, 1 volt each. We find that as the drain-source voltage gets increased, the drain attracts more and more electrons, and I_D increases correspondingly. The increase in I_D results in the increase in the reverse bias between the gate and the drain. We find that the depletion layer between gate and drain gets wider and wider as the reverse bias is increased.

4.3.3 Drain Characteristics

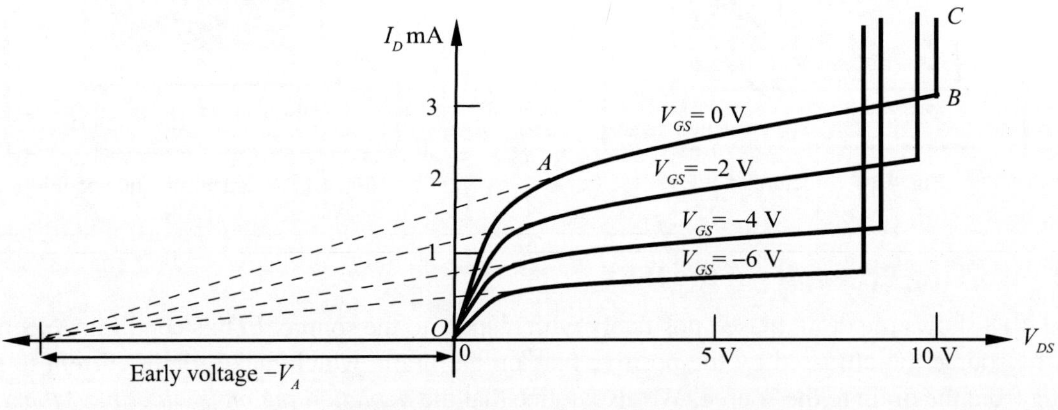

Fig. 4.19 Drain characteristics of JFET.

Figure 4.19 shows the drain characteristics, a plot between drain-source voltage V_{DS} and drain current I_D, with the gate-source voltage V_{GS} kept constant. Initially, we fix V_{GS} at 0 V, vary V_{DS} in steps of 1 volt each, as stated above, and note the corresponding values of I_D. The resulting values are tabulated, and using the tabulated values, the characteristics for $V_{GS} = 0$ is plotted, as shown in Fig. 4.19.

Consider the region *OA* in the characteristic. We find that I_D increases linearly with V_{DS}. This is due to the fact that as V_{DS} is increased more and more electrons get attracted by the drain resulting in an increase in I_D. We also notice that the increase in I_D results in an increase in the depletion width between drain and gate. The increase in I_D continues until the point *A* is reached, at which point, all the electrons emitted from the source are attracted by the drain and I_D reaches its maximum value.

At point *A*, the maximum value of I_D produces maximum drop in the channel, which in turn produces maximum reverse bias between the drain and gate terminals. This results in the channel getting almost **pinched-off**.

At pinch-off, the channel cannot completely close; for, if the channel closes, the flow of I_D would stop completely. But, if the flow of I_D stops completely, then the entire depletion layer will disappear. This will open up the channel and I_D will flow again. This confirms the theory that the channel cannot get completely pinched-off.

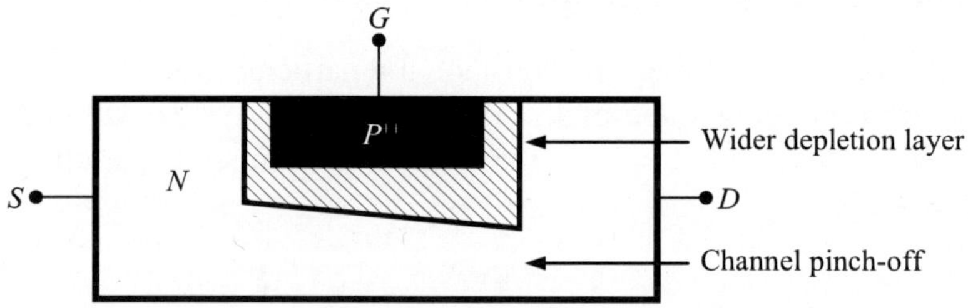

Fig. 4.20 Channel at pinch-off.

I_D is a maximum at pinch-off, as shown in Fig. 4.20. This is similar to the situation in a river where a constriction in the path blocks the flow of water. We notice that to maximise the current flow through a constriction, the velocity of the flow has to be increased proportionately. In the case of the FET also, this is true. Thus, we find that at pinch off the velocity of the electrons is increased to maximise the drain-current flow. This drain current is called the **saturation drain current** and is designated as I_{DSS}.

In the region *AB* of the characteristics shown in Fig. 4.20, the current remains more or less constant. This is because at point *A* itself, all the electrons emitted from the source would have reached the drain, and current becomes a maximum. Therefore, even if V_{DS} is increased I_D cannot increase.

Now, we refer to the point *B* in the curve. At this point, the reverse voltage is so high that avalanche breakdown will occur. At this point, the current suddenly rises to a very high value, which unless is limited by a suitable resistor, will destroy the device. The breakdown region is marked as region *BC* in Fig. 4.20.

Let us now fix V_{GS} at −2 V. The experiment is repeated by varying V_{DS} and noting the corresponding values of I_D, as before. By applying an external reverse bias between gate and source, a depletion region is created between drain and gate. This depletion region is dependent only on the gate-supply voltage and is totally *independent* of the drain-current flow.

Now, if we apply V_{DD}, as before I_D would start flowing again, and the whole story described above repeats. We now find that the V_{DS}-I_D curve for $V_{GS} = -2$ V is similar to, and almost parallel to that for $V_{GS} = 0$ V. However, the current in this case is less than that in the previous case. This is because of the depletion region already formed as a result of the reverse bias between the gate and the source, which reduces the channel width.

The experiment is repeated for various values of V_{GS}. The results are tabulated and the characteristics are drawn as shown in Fig. 4.20. We find that, as V_{GS} is increased in the negative direction, i.e., −2 V, −4 V, etc., the drain current gets reduced proportionately. Ultimately, at a large negative value of V_{GS}, the channel will be completely cut-off or closed so that no drain current can flow. This is called the **cut-off** condition. *Complete channel cut-off is possible because, in this case, it is the external gate-source voltage that brings in the channel cut-off, and not by the drain current as in the case where* V_{GS} *is zero.*

4.3.4 Mutual or Transfer Characteristic

The transfer curve is plotted by keeping V_{DS} constant at a particular value, say 5 V, and by varying V_{GS} and noting the corresponding value of I_D. As shown in Fig. 4.21, we find that the curve is almost linear. We also find that at $V_{GS} = -V_P$, the channel is completely cut-off or pinched-off. For a typical JFET, $V_P = -8$ V.

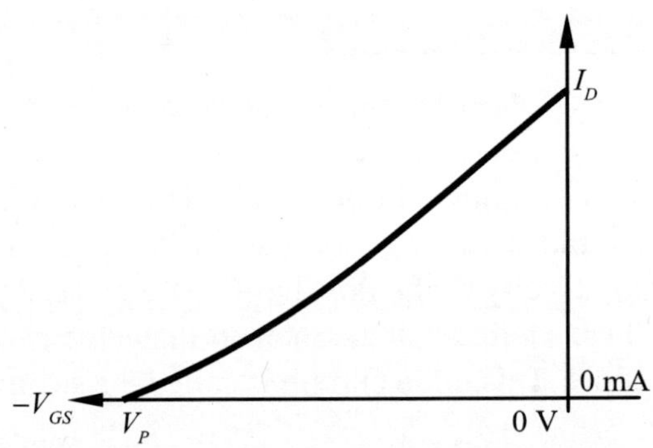

Fig. 4.21 Transfer or mutual characteristic of JFET.

The slope of the transfer curve is called the mutual conductance and is defined as

$$g_m = \left[\frac{\Delta I_D}{\Delta V_{GS}}\right]_{V_{DS}=\text{constant}} \tag{4.30}$$

4.3.5 Amplification Characteristic

We also can plot the amplification characteristic of the JFET, which is the plot between V_{DS} and $-V_{GS}$, with I_D kept constant. This is shown in Fig. 4.22. The slope of the characteristic is known as the **amplification factor μ** of JFET. Mathematically, amplification factor is defined as:

$$\mu = -\left[\frac{\Delta V_{DS}}{\Delta V_{GS}}\right]_{I_D = \text{constant}} \tag{4.31}$$

The negative sign in the equation indicates that the variation in V_{DS} is opposite to the variation in V_{GS}. Amplification factor represents the maximum amount of amplification that the device is capable of producing.

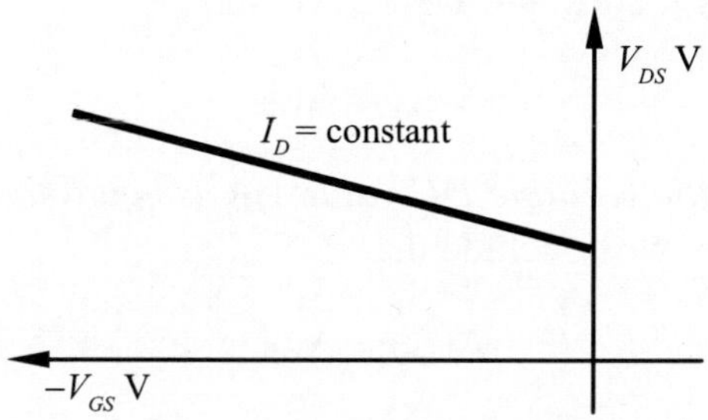

Fig. 4.22 Amplification characteristic.

4.3.6 Relation among μ, g_m, and r_d

From the definitions, we find that there exists a definite relation among the JFET parameters μ, g_m and r_d. From experiments, we find that

$$i_D = f(v_{DS}, v_{GS}) \tag{4.32}$$

where we have used instantaneous total values of drain current, drain-source voltage, and gate-source voltage. By instantaneous total value, we mean the variable DC value. As an example, we express

$$v_{DS} = V_D + v_{ds} \tag{4.33}$$

where v_{DS} = variable DC, V_{DS} = pure DC, and v_d = pure AC. Using Taylor's expansion, Eq. (4.32) may be written as

$$\Delta i_D = \left(\frac{\partial i_D}{\partial v_{DS}}\right)\Delta v_{DS} + \left(\frac{\partial i_D}{\partial v_{GS}}\right)\Delta v_{GS} \tag{4.34}$$

where Δv_{DS} = a small change in v_{DS}, Δv_{GS} = a small change in v_{GS}, and Δi_D = corresponding change in i_D. Equation (4.34) suggests that the total variation in the value of the drain current i_D is the equal to the sum of the variations in i_D due to the variations in v_{DS} and v_{GS}. We now define the following terms:

$$\frac{\partial i_D}{\partial v_{DS}}\Big|_{\Delta v_{GS}=0} = r_d \text{ (drain resistance)} \tag{4.35}$$

$$\frac{\partial i_D}{\partial v_{GS}}\Big|_{\Delta v_{DS}=0} = g_m \text{ (transfer conductance)} \tag{4.36}$$

Using the above in Eq. (4.34), we get

$$\Delta i_{\text{D}} = (1/r_{\text{d}})\,\Delta v_{\text{DS}} + g_{\text{m}}\Delta v_{\text{GS}} \tag{4.37}$$

Now, differentiating Eq. (4.33), we get

$$\Delta v_{\text{DS}} = \Delta V_{\text{D}} + \Delta v_{\text{ds}} \tag{4.38}$$

In Eq. (438), $\Delta V_{\text{D}} = 0$ since V_{D} = constant. Hence we find

$$\Delta v_{\text{DS}} = \Delta v_{\text{ds}} \tag{4.39}$$

This means that *the change in the variable DC value* v_{DS} *is equal to the variation in the AC value* v_{ds}. Since v_{ds} is a variable value, we conclude that

$$\Delta v_{\text{DS}} = \Delta v_{\text{ds}} \tag{4.40}$$

which is the AC value in the waveform. In a similar fashion, we get that

$$\Delta v_{\text{GS}} = v_{\text{gs}} \tag{4.41}$$

$$\Delta i_{\text{D}} = i_{\text{d}} \tag{4.42}$$

Substituting the new values, Eq. (4.37) becomes

$$i_{\text{d}} = (1/r_{\text{d}})\,v_{\text{ds}} + g_{\text{m}}r_{\text{d}} \tag{4.43}$$

This equation is used to develop the small-signal equivalent circuit of the JFET, shown in Fig. 4.23.

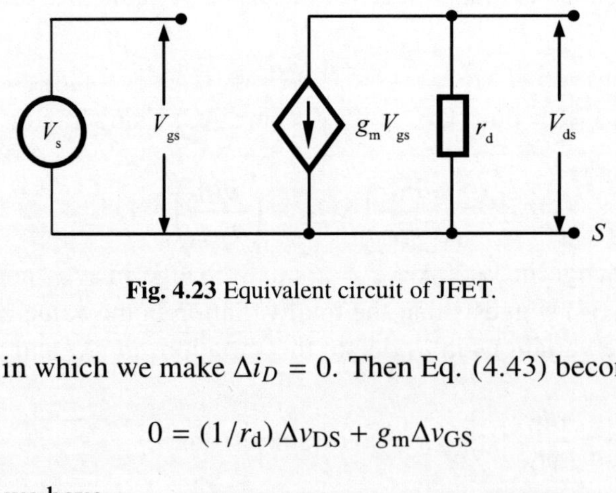

Fig. 4.23 Equivalent circuit of JFET.

Consider Eq. (4.43) in which we make $\Delta i_D = 0$. Then Eq. (4.43) becomes

$$0 = (1/r_{\text{d}})\,\Delta v_{\text{DS}} + g_{\text{m}}\Delta v_{\text{GS}} \tag{4.44}$$

Rearranging Eq. (4.44), we have

$$(1/r_{\text{d}})\,\Delta v_{\text{DS}} = -g_{\text{m}}\Delta v_{\text{GS}} \tag{4.45}$$

From Eq. (4.45), we get

$$-\frac{\Delta v_{DS}}{\Delta v_{GS}} = g_m \times r_d \tag{4.46}$$

But, by definition, we have

$$-\frac{\Delta v_{DS}}{\Delta v_{GS}} = \mu \tag{4.47}$$

Combining Eqs. (4.46) and (4.47), we find that

$$\mu = g_m r_d \tag{4.48}$$

ONE-WORD/MULTIPLE-CHOICE QUESTIONS

M4.1 A field-effect transistor works on the principle of ________ field controlling drain current.

M4.2 The situation of drain current becoming zero with large negative gate voltage is called ________.

M4.3 The situation of drain current becoming just saturated is called ________.

M4.4 The drain current under saturation is called ________ drain current.

M4.5 The slope of the drain characteristic is called ________.

M4.6 The slope of the mutual characteristic is called ________.

M4.7 The slope of the amplification characteristic is called ________.

M4.8 Amplification factor is defined as ________.

M4.9 Transconductance is defined as ________.

M4.10 Drain resistance is defined as ________.

M4.11 The channel enhanced in a MOSFET is known as ________ layer.

M4.12 A four-layer *PNPN* diode is called the ________ diode.

M4.13 The anode current in a four-layer *PNPN* diode is given by ________.

M4.14 The anode current in a four-layer *PNPN* diode, when $\alpha_p = 1 = \alpha_n$ is ________.

M4.15 The state $I_A = I_{CO}$ is called the forward ________.

M4.16 The state $I_A = \infty$ is called the forward ________ state.

M4.17 The current reached by the device after the device has broken down in the forward state is called the ________ current.

M4.18 The turn-off current is called the ________ current.

M4.19 An SCR has a ________ resistance region in its characteristic.

M4.20 An SCR is a ________ device.

M4.21 The factor η of a UJT s known as ________ ratio.

M4.22 The oscillator constructed using a UJT is known as ________ oscillator.

M4.23 The effect with which channel resistance gets varied in a UJT is known as ________.

M4.24 The maximum point in the *VI* characteristic of a UJT is known as ________.

M4.25 The minimum point in the *VI* characteristic of a UJT is known as ________ point.

REVIEW QUESTIONS

1. Describe the constructional features of the four-layer Shockley diode.
2. Explain the working principles of the four-layer diode.
3. Analyse the operation of the four-layer diode using the two-transistor analogy.
4. What is the forward-blocking state of an SCR?
5. Explain the *VI* characteristic of the Schockley diode.
6. Explain the *VI* characteristic of SCR.
7. What are the various triggering methods of SCR?
8. Explain the constructional features of the uni-junction transistor.
9. Draw the equivalent circuit of a UJT.
10. Explain the working principles of UJT.
11. Describe the *VI* characteristic of UJT.
12. Explain operation of a UJT relaxation oscillator.
13. Derive the equation for frequency of oscillation of UJT relaxation oscillator.
14. Explain how the UJT acts as a triggering device for SCR.

15. With suitable sketches, explain the structure of a typical JFET.
16. Explain the working principles of JFET.
17. What is the pinch-off?
18. What is cut-off?
19. Explain the difference between pinch-off and cut-off.
20. Explain the drain characteristics of JFET.
21. What is Early voltage?
22. Draw the mutual or transfer characteristic of JFET, and discuss the nature of the curves.
23. Deduce the relation among μ, g_m, and r_d of a JFET.
24. Explain the amplification characteristics of JFET.
25. Derive the expression for the small-signal operation of a JFET.
26. Draw the small-signal equivalent circuit of JFET.

UNSOLVED PROBLEMS

P4.1 Compute the values of small-signal parameters of a junction field-effect transistor using the data given in the following table.

V_{GS}	V_{DS}	I_D
−2	0	0
	2	0.5
	4	1
	6	2
	8	3
	10	4
−3	0	0
	2	0
	4	0.5
	6	1
	8	2
	10	3

P4.2 Compute the value of amplification factor of a junction field-effect transistor if at I_D = 2 mA, when V_{DS} = 15 V, V_{DS} = −1 V and when V_{DS} = 5 V, V_{GS} = 0 V.

P4.3 Compute the value of transconductance of a junction field-effect transistor, if at V_{DS} = 10 V, when $V_{GS} = -2$ V, $I_D = 4$ mA and when $V_{GS} = -3$ V, $I_D = 2$ mA.

P4.4 Compute the value of drain resistance of a junction field-effect transistor, if at $V_{GS} = -3$ V, when $V_{DS} = 15$ V, $I_D = 4$ mA and when $V_{DS} = 10$ V, $I_D = 2$ mA.

P4.5 Find the frequency of oscillation of a relaxation oscillator using uni-junction transistor given that η of UJT is 0.7, $R = 6k\Omega$, and $C = 0.01$ μF.

P4.6 Find the frequency of oscillation of a relaxation oscillator using uni-junction transistor given that R_{B1} of UJT is 6 kΩ, $R_{B2} = 4$ k Ω, and $C = 0.001$ μF.

P4.7 Compute the value of the anode current in the case of an SCR under switch-on condition, if the respective current gains are $\alpha_p = \alpha_n = 0.498$.

P4.8 Compute the value of the anode current in the case of an SCR under switch-on condition, if the respective current gains are $\alpha_p = \alpha_n = 0.4995$.

CHAPTER 5

Amplifiers and Oscillators

5.1 DECIBEL AND HALF-POWER POINTS

5.1.1 Bel and Decibel

Figure 5.1 shows the block schematics of an *R-C* coupled amplifier. Let the power input P_i to this amplifier be 1 mW and let the power output from the amplifier P_o be 100 mW. The power gain A_p of this amplifier can be computed as

$$A_p = P_o/P_i = 100/1 = 100 \tag{5.1}$$

Equation (5.1) has no units attached to it. Now, taking common logarithm of both sides of Eq. (5.1), (ignoring the numerical part for the moment) we find

$$A_p \text{ in bels} = \log(P_o/P_i) \tag{5.2}$$

In Equation (5.2), we have attached the unit of bel to the power ratio. Using the numerical value in Eq. (5.1), we get the power gain in the given example as

$$A_p = \log 100 = 2 \text{ bels} \tag{5.3}$$

Equation (5.3) states that the power gain of the given amplifier is one bel. Since this is a large unit, we use a sub-multiple of this unit: the decibel (abbreviated as dB). As per basic definition of units,

$$1 \text{ bel} = 10 \text{ decibels} \tag{5.4}$$

Using this definition, we find that the power gain of the amplifier

$$A_p = \log 100 = 2 \text{ B} = 20 \text{ dB} \tag{5.5}$$

Caution: It should be clearly understood that

$$1 \text{ bel} \neq \log(P_o/P_i)$$

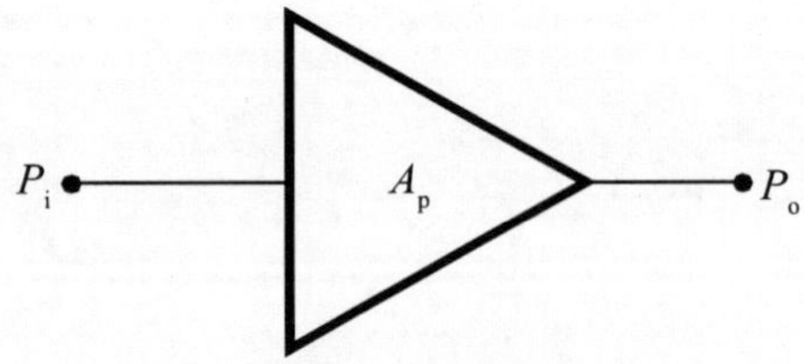

Fig. 5.1 Amplifier.

Instead, we state that the power ratio P_o/P_i has *so many* bels in it. Similarly

$$1 \text{ decibel} \neq 10 \log (P_o/P_i)$$

But the power ratio P_o/P_i has $10 \times$ so many bels = so many decibels in it.

From the definitions, we find that bel and decibel are logarithmic units. Hence, we should be careful in interpreting the power ratios expressed in these units. For example, 30 dB and 40 dB represent power ratios of 1000 and 10,000, respectively.

It is customary to redefine bels and decibels in terms of voltage and current ratios. For this, we use the relations (assuming $R = 1$).

$$P = V^2/R = I^2R = V^2 = I^2 \tag{5.6}$$

Now, using the relations given in Eq. (5.6), we re-express the power ratios in the form

$$A_p = P_o/P_i = (V_o/V_i)^2 \tag{5.7}$$

In Eq. (5.7), using the definition of bel and decibel, the expression for power gain becomes

$$A_p = 10 \log (V_o/V_i)^2 = 20 \log (V_o/V_i) \text{ dB} \tag{5.8}$$

Equation (5.8) is the most common form of representation of gain in decibels. The following numerical examples will illustrate this concept in depth.

Example 5.1 Obtain the gain in terms of decibels of the following.

(a) $P_i = 1\ \mu$W and $P_o = 1000\ \mu$W
(b) $P_i = 1\ \mu$W and $P_o = 100$ mW
(c) $V_i = 1\ \mu$V and $P_o = 100\ \mu$V
(d) $V_i = 1\ \mu$V and $P_o = 1000$ mV

Solution

(a) Gain = $10 \log(1000/1) = 30$ dB
(b) Gain = $10 \log(100 \times 10^3/1) = 50$ dB (in this case, we have used 1 mW = 1000 μW)
(c) Gain = $20 \log(100/1) = 40$ dB
(d) Gain = $20 \log(1000 \times 10^3/1) = 120$ dB

5.1.2 Half-Power Points

Figure 5.2 shows the frequency-response curve of an *R-C* coupled amplifier. Frequency-response curve of an amplifier is the frequency-vs-gain plot of the amplifier.

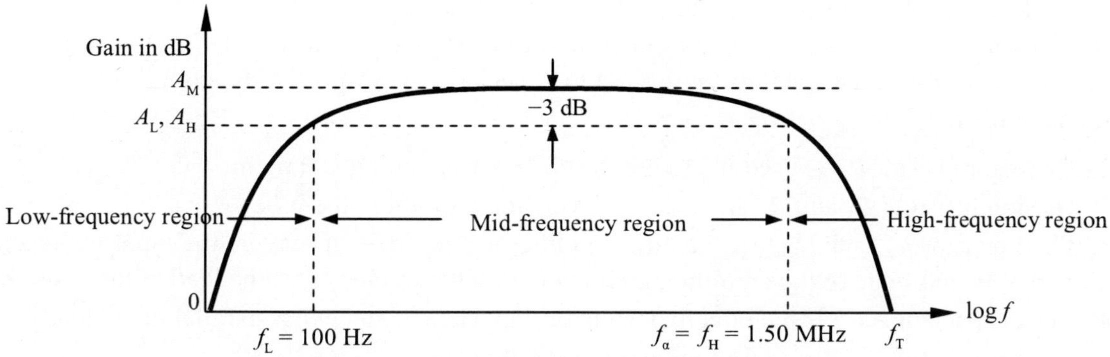

Fig. 5.2 Frequency-response plot of an *RC*-coupled *CE* amplifier.

It can be seen from Fig. 5.2 that that the gain is constant in the mid-frequency region and drops to zero in the high and low-frequency regions. The reasons for these will be explained later. Our present task is to define the lower and upper half-power points and the bandwidth of the amplifier.

It is quite natural to assume that an amplifier must output a constant power for all the frequencies of its operation. However, due to the action of stray capacitances present in its circuit, no amplifier can give constant output for all the frequencies of operation. Since there is reduction in gain in the low- and high-frequency regions as shown in Fig. 5.2, we assume that the amplifier is useful for all the frequencies, which delivers or more of output power.

We call the lower and upper frequencies at which the output power is 50% of the mid-band output power as the *lower* and *upper half-power frequencies*. Thus at the lower half-power frequencies f_L and at the upper half-power frequencies f_H, the power gains may be expressed as

$$A_{PL} = \frac{A_{PM}}{2} = A_{PH} \tag{5.9}$$

Taking the power gain ratios and expressing them in dB, we get

$$10 \log \frac{A_{PM}}{A_{PL}} = 10 \log \frac{A_{PM}}{A_{PH}} = 10 \log 2 = 3 \text{ dB} \tag{5.10}$$

From Eq. (5.10), we observe that the power gain of the amplifier at the lower and upper half-power frequencies are down by 3 dB from its power gain in the mid-band region. We define the bandwidth of an amplifier BW as

$$\text{BW} = f_H - f_L \tag{5.11}$$

Since this bandwidth is obtained by a 3-dB reduction in gain, it is customary to call it as the 3-dB *bandwidth*.

5.2 SINGLE-STAGE CE AMPLFIER

Figure 5.3 shows a single-stage *R-C* coupled class-A voltage amplifier using BJT. The second stage is represented by the parallel combination of load resistor R'_L and load capacitor, which are, respectively, the input impedance and capacitance of that stage.

As shown in Fig. 5.3, the first stage consists of bipolar junction transistor T as the amplifying device. The resistors and capacitors connected to T perform the respective functions noted against each one of them.

1. Collector resistor R_C is used for fixing the collector operating (Q) point.
2. Biasing resistors R_1 and R_2 are used for fixing the Q-point in the base.
3. Load resistor R'_L is the actual load the amplifier has to drive. It may be the input impedance of the second stage of the amplifier, and/or any resistor connected as the load of the amplifier.
4. Load capacitance C'_L is the effective parasitic capacitance that exists parallel to the load.
5. Emitter resistor R_E is used for temperature stabilisation.
6. Bypass capacitor C_E is used for bypassing emitter resistor R_E and thus reduce the effect of negative feedback.
7. Coupling capacitors C_{C1} and C_{C2} are used for DC blocking between stages to prevent level-shifting problems.

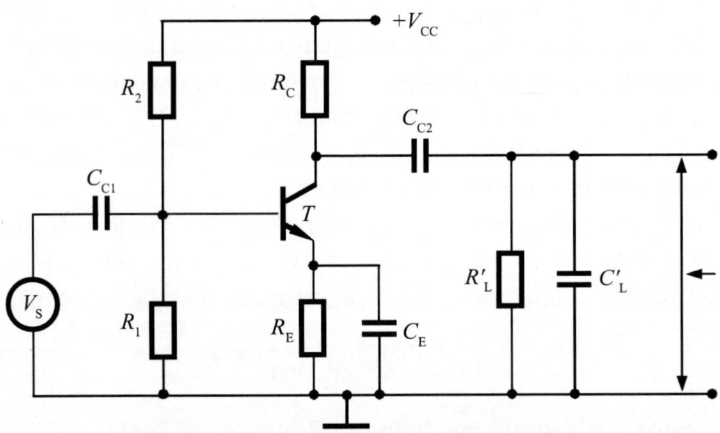

Fig. 5.3 *RC*-coupled class-A voltage amplifier.

5.2.1 Biasing Voltages and Currents in Single-Stage CE Amplifier

Figure 5.4 shows the voltage drops at various terminals of a properly designed *CE* amplifier. Assume that the supply voltage $V_{CC} = 10$ V, and the collector current is $I_C = 1$ mA.

Now, assuming that the voltage drop across emitter resistor R_E for temperature stability is 0.1 V_{CC}, the optimum output voltage can be computed as

$$V_{CE} = V_{RE} = (V_{CC} - 0.1V_{CC})/2 \tag{5.12}$$

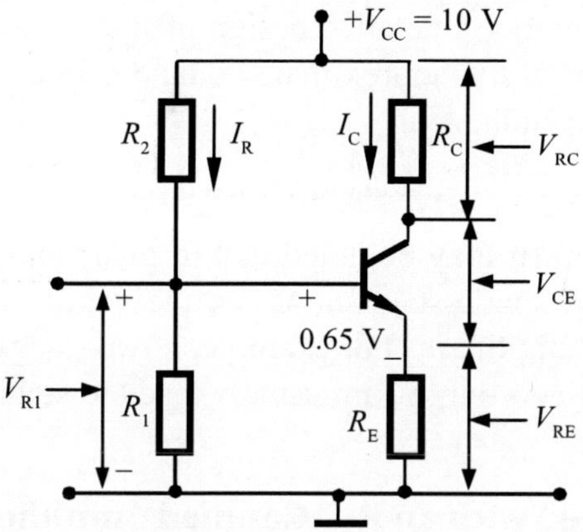

Fig. 5.4 Voltage distributions.

With V_{CC} = 10 V, we obtain

$$V_{CE} = V_{RE} = (10 - 1)/2 = 4.5 \text{ V}$$

Then, since I_C = 1 mA, we find the value of the collector resistor as

$$R_C = V_{RC}/I_C = 4.5 \text{ V}/1 \text{ mA} = 4.5 \text{ k}\Omega \tag{5.13}$$

Similarly, we obtain the value of the emitter resistor as

$$R_E = V_{RE}/I_C = 1 \text{ V}/1 \text{ mA} = 1 \text{ k}\Omega \tag{5.14}$$

Now, to find the values of base biasing resistors and R_2, we first find the voltage drop across resistor R_1 by using the relation

$$V_{R1} = V_{RE} + V_{BEQ} = \frac{V_{CC}R_1}{R_1 + R_2} 1 + 0.65 = 1.65 \text{ V} \tag{5.15}$$

where we have assumed the base biasing voltage V_{BEQ} as 0.65 V, which is the average value of the cut-in voltage of 0.5 V and saturation voltage of 0.8 V (see Fig. 5.5). Assuming a DC current gain (β or h_{FE}) of 100, we find the base current

$$I_B = I_C/\beta = 1 \text{ mA}/100 = 10 \text{ }\mu\text{A} \tag{5.16}$$

From analysis, for temperature stability, we assume that base equivalent resistance

$$R_{BB} = \frac{R_1R_2}{R_1 + R_2} = \frac{\beta}{10} = 100/10 = 10 \text{ k}\Omega \tag{5.17}$$

Solving Eqs. (5.15) and (5.17), we obtain

$$R_1 \approx 12 \text{ k}\Omega \text{ and } R_2 \approx 60 \text{ k}\Omega$$

This completes the basic design. For a detailed design of the *CE* amplifier, the reader is directed to Appendix A. The expression for the approximate voltage gain of the amplifier using the hybrid analysis may be found from Appendix A as

$$A_v \approx h_{fe}R_L/(R_S + h_{ie}) \tag{5.18}$$

where h_{fe} is the AC current gain (it may be noted that in many cases, $h_{fe} = h_{FE}$). It may also be noted that for a typical BJT *CE* single-stage amplifier, the voltage gain is about 100 to 200. For larger gain, we use multi-stage amplifiers. For example, a two- stage amplifier may have a typical gain of 10^4 to 10^5. However, a very large gain usually leads to stability problems in the amplifier system.

5.2.2 Waveforms Associated with an R-C Coupled Amplifier

Figure 5.5 shows the input and output waveforms for various biasing conditions of an *R-C* coupled amplifier. In Fig. 5.5, we find the *Q*-point to be in the middle of the active region. In this case, we have assumed $V_{BE\,min} = 0.5$ V, and $V_{BE\,max} = 0.8$ V, which are typical values for silicon transistors.

For an ideal bipolar junction transistor, assuming exponential transfer characteristics, the *Q*-point will be at 0.7 V. However, for an Indian transistor, it is found that the transfer characteristic is more linear, and therefore we can see that the *Q*-point lies between 0.62 V to 0.65 V. In our present discussion, we have chosen it to be 0.65 V, as shown in Fig. 5.5. It can be seen that in this case the output waveforms (I_B and I_C) give maximum swing and exhibit no distortion. It may be noted that, since both the waves are similar in shape, except for the difference in their magnitudes ($I_C = \beta I_B$), they are drawn as a single curve, as shown in Fig. 5.6.

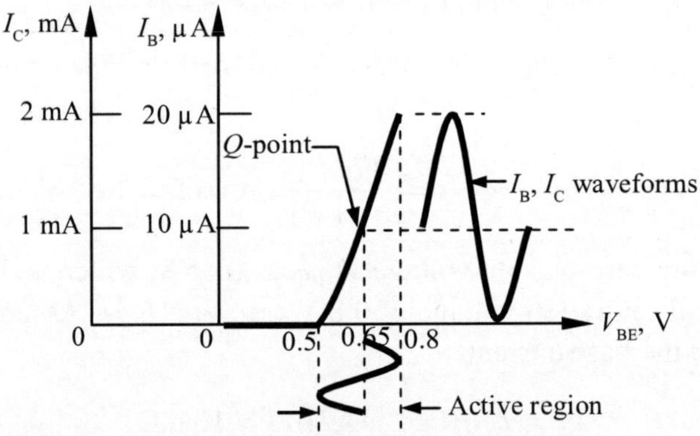

Fig. 5.5 *Q*-point in the middle of active region.

In Fig. 5.6, the *Q*-point has been shifted to 0.7 V (nearer to $V_{BE\,max}$ of 0.8 V) because of improper biasing. In this case, the output swing is small as the input swing is small. If the input swing exceeds the limits, then distortion will take place, as shown in Fig. 5.6. Similarly, as shown in Fig. 5.7, the *Q*-point is now shifted to 0.6 V (nearer to $V_{BE\,min}$ of 0.5 V), again due to improper biasing. In this case also we observe that the output swing is small due to small input swing, and distortion will take place if the input swing exceeds the limits, as shown in Fig. 5.7.

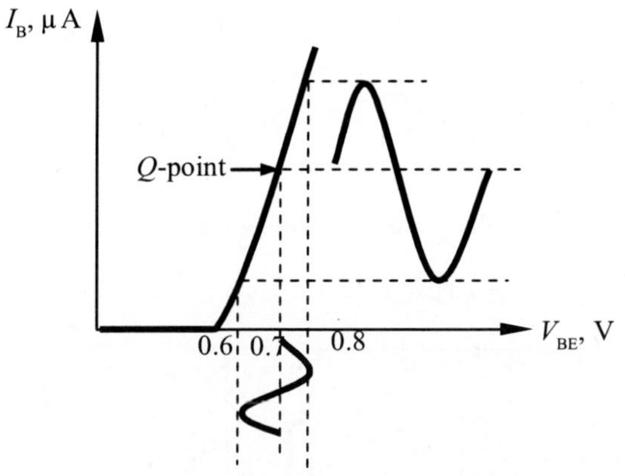

Fig. 5.6 Q-point nearer to V_{BEmax}.

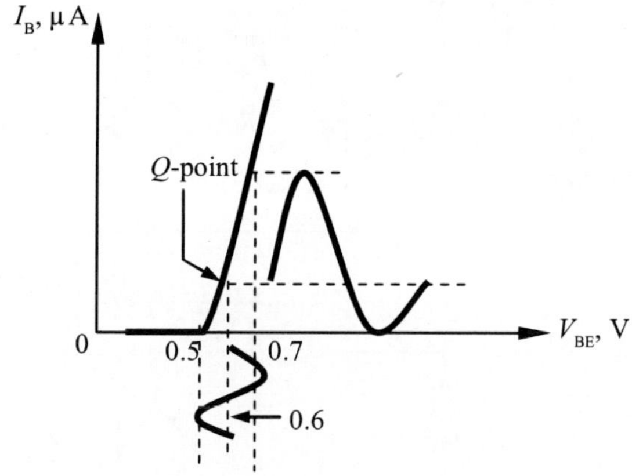

Fig. 5.7 Q-point nearer to V_{BEmin}.

In Fig. 5.8, the Q-point has been shifted again to 0.7 V; but the input swing is large. This makes the output distorted, as can be seen from the clipped output wave. In a similar way, when the input swing is large and the Q-point shifted again to 0.5, then the output can be found to be distorted as shown in Fig. 5.9.

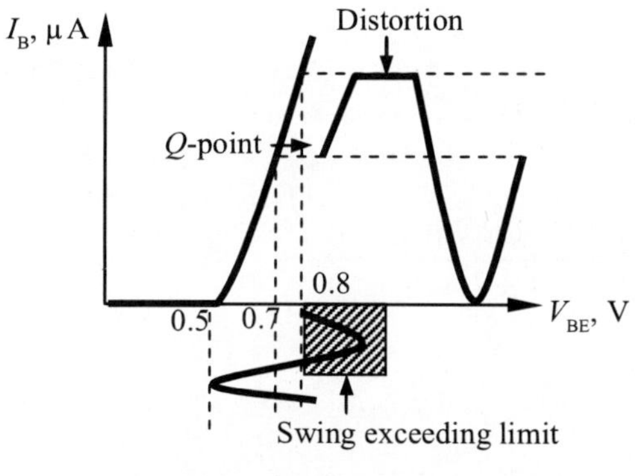

Fig. 5.8 Q-point nearer to V_{BEmax}.

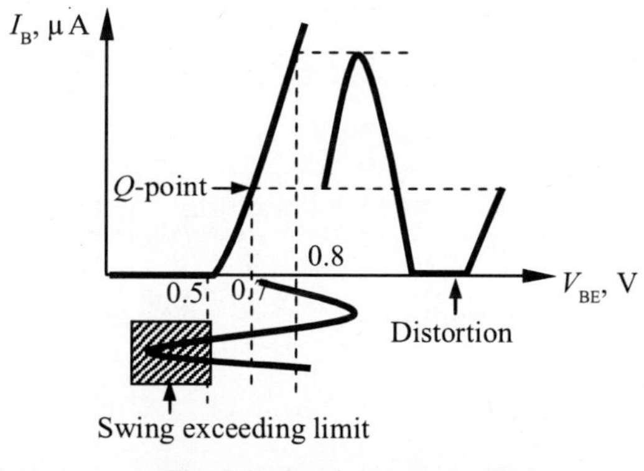

Fig. 5.9 Q-point nearer to V_{BEmin}.

Thus we conclude that the output swing will be symmetrical and maximum only when the Q-point is exactly in the middle of the active region of the VI characteristics of the amplifier. Shifting of the Q-point from the middle portion will lead to the distortion in the output waveform.

5.3 ANALYSIS OF R-C COUPLED C-E AMPLIFIER USING h PARAMETERS

The circuit shown in Fig. 5.10(a) is reduced to the simplified form shown in Fig. 5.10(b). In the simplified equivalent circuit, we have again decided that the parallel combination of R_C and R'_L can be reduced to an effective load resistance R_L. However, there is one more resistance appearing in the equivalent circuit shown in Fig. 5.10(b), and this is the output impedance $1/h_{oe}$ of BJT.

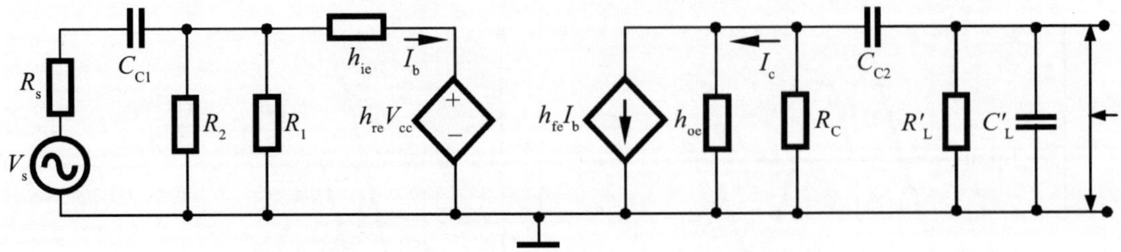

Fig. 5.10(a) Hybrid equivalent circuit of BJT (*CEC*).

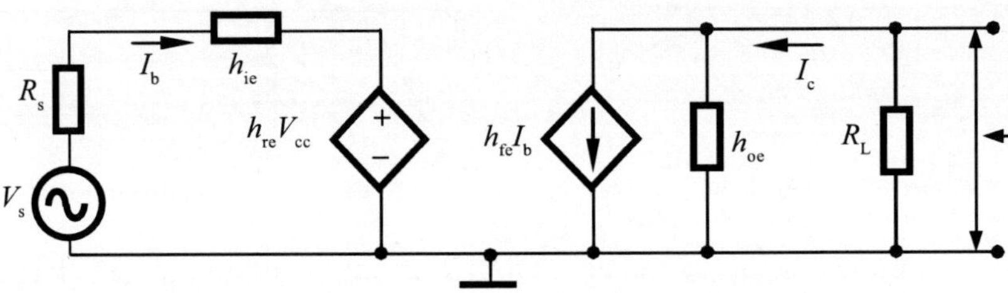

Fig. 5.10(b) Reduced and simplified hybrid equivalent circuit.

5.3.1 Expression for Voltage Gain

From the output loop of Fig. 5.10(b), we get

$$V_o = -h_{fe}I_b\,[(1/h_{oe})\,||(R_L)] \tag{5.19}$$

Rearranging

$$V_o = -\beta I_b\left[\frac{(1/h_{oe}\,)\,R_L}{(1/h_{oe}\,)+R_L}\right] = -\frac{\beta I_b R_L}{1+h_{oe}R_L} \tag{5.20}$$

Now, from the input loop

$$V_s = I_b\,(R_s + h_{ie}) + h_{re}V_o \tag{5.21}$$

From Eq. (5.20)

$$I_b = -\frac{V_o\,(1+h_{oe}R_L)}{\beta R_L} \tag{5.22}$$

Substituting this value in Eq. (5.21) yields

$$\begin{aligned} V_s &= -\frac{V_o\,(1+h_{oe}R_L)\,(R_s+h_{ie})}{\beta R_L} + h_{re}V_o \\ &= \frac{V_o\,[(1+h_{oe}R_L)\,(R_s+h_{ie}) - \beta R_L h_{re}]}{-\beta R_L} \end{aligned} \tag{5.23}$$

Rearranging Eq. (5.23), we get the expression for the voltage gain as

$$A_v = \frac{-\beta R_L}{(1+h_{oe}R_L)\,(R_s+h_{ie}) - \beta R_L h_{re}} \tag{5.24}$$

Since $h_{re}(\approx 10^4)$ and $h_{oe}(\approx 10^4)$ are numerically extremely small quantities compared with other terms in the expression, we may neglect them as well as the terms associated with them, and this reduces Eq. (5.24) to

$$A_v = \frac{-\beta R_L}{R_s + h_{ie}} \tag{5.25}$$

5.3.2 Expression for Current Gain

Consider Eq. (5.20) again, which states that

$$V_o = -\frac{\beta I_b R_L}{1 + h_{oe} R_L}$$

But from the output loop, shown in Fig. 5.9(b), we have

$$V_o = -I_c R_L \tag{5.26}$$

Then Eq. (5.20) becomes

$$-I_o R_L = -\frac{\beta I_b R_L}{1 + h_{oe} R_L} \tag{5.27}$$

Reducing and rearranging Eq. (5.27) yields current gain

$$A_i = \frac{I_o}{I_b} = \frac{\beta I}{1 + h_{oe} R_L} \tag{5.28}$$

As before, since h_{oe} is negligible, we get

$$A_i = \beta \tag{5.29}$$

5.3.3 Expression for Power Gain

$$A_p = A_v \times A_i = \beta^2 R_L/(R_s + h_{ie}) \tag{5.30}$$

5.3.4 Expression for Input Impedance

Consider Eq. (5.21), which states that

$$V_s = I_b\,(R_s + h_{ie}) + h_{re} V_o$$

Substituting for V_o from Eq. (5.19) into Eq. (5.20), we get

$$V_s = I_b\,(R_s + h_{ie}) - h_{re}\left(\frac{\beta I_b R_L}{1 + h_{oe} R_L}\right) \tag{5.31}$$

Rearranging and simplifying Eq. (5.31) yields the expression for the input impedance

$$Z_i = \frac{V_s}{I_b} = (R_s + h_{ie}) - \frac{\beta h_{re} R_L}{1 + h_{oe} R_L} \tag{5.32}$$

Neglecting small terms h_{oe} and h_{re} in Eq. (5.32), we find

$$Z_i = R_s + h_{ie} \tag{5.33}$$

5.3.5 Expression for Output Impedance

To determine the output impedance, we redraw the equivalent circuit as shown in Fig. 5.11.

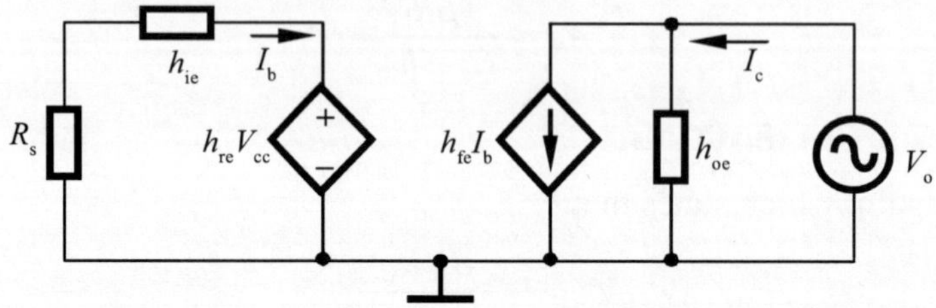

Fig. 5.11 Hybrid equivalent circuit for determining the output impedance.

From the output loop, we find

$$I_o = \beta I_b + h_{oe}V_o \tag{5.34}$$

From the input loop, we have

$$I_b = -h_{re}V_o/h_{ie} \tag{5.35}$$

Using Eqs. (5.34) and (5.35), we get

$$I_o = -\beta\frac{h_{re}V_o}{h_{ie}} + h_{oe}V_o = V_o\left(-\frac{\beta h_{re}}{h_{ie}} + h_{oe}\right) \tag{5.36}$$

Thus, output admittance

$$Y_o = \frac{I_o}{V_o} = -\frac{\beta h_{re}}{h_{ie}} + h_{oe} \tag{5.37}$$

If h_{re} is neglected Eq. (5.37) reduces to

$$Y_o = h_{oe} \tag{5.38}$$

5.4 TWO-STAGE R-C COUPLED C-E AMPLIFIER

We can construct a two-stage *CE* amplifier by connecting two identical single-stage amplifiers in cascade, as shown in Fig. 5.12.

It can be seen that the gain of the second stage is given by

$$A_{v2} = \frac{-\beta_2 R_{L2}}{h_{ie2}} \tag{5.39}$$

where the subscripts, viz., 2, represent the respective values related to the second stage. It may be noted that the effective load resistance of the second stage is actually the parallel combination of the collector resistance and the actual load resistance of that stage. That is

$$R_{L2} = R_{C2}\|\, R_L \tag{5.40}$$

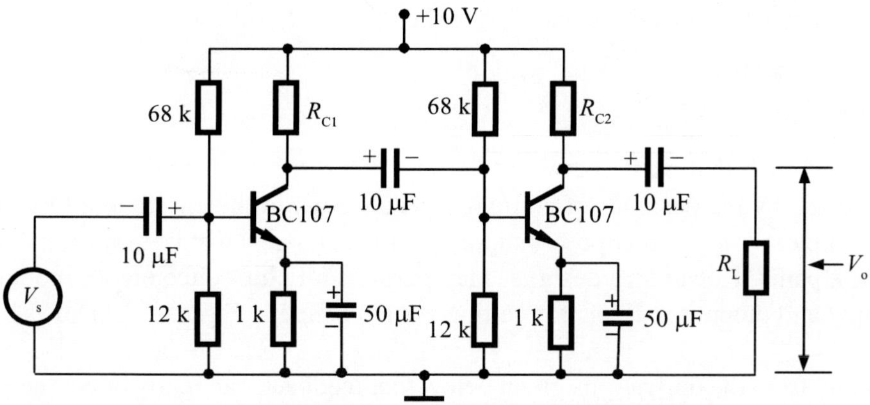

Fig. 5.12 Two-stage voltage amplifier.

The gain of the first stage is similarly given by

$$A_{v1} = -\beta_1 R_{L1}/h_{ie1} \tag{5.41}$$

where the subscripts 1 the respective values related to the first stage. It may be noted that the effective load resistance of the first stage is actually the parallel combination of the collector resistance of the first stage and the input impedance of the second stage. That is

$$R_{L1} = R_{C1} \parallel h_{ie2} \tag{5.42}$$

Thus the overall voltage gain of the two stages is given by

$$A_v = A_{v1}A_{v2} = \frac{-\beta_1 R_{L2}}{h_{ie1}} \times \frac{-\beta_2 R_{L2}}{h_{ie2}} = \frac{\beta^2 R_{L2} R_{L1}}{h_{ie1} h_{ie2}} \tag{5.43}$$

where we have assumed that the transistors are identical and hence $\beta_1 = \beta_2 = \beta$. Usually $R_{C1} >> h_{ie2}$ so that $R_{L1} = h_{ie2}$. Using this relation, Eq. (5.43) gets simplified as

$$A_v = \beta^2 R_{L2}/h_{ie1} \tag{5.44}$$

5.5 AMPLIFIERS WITH NEGATIVE FEED BACK

5.5.1 Introduction

Amplifiers in general are classified as:

- Those without feedback connections
- Those with feedback connections

Amplifiers with feedback connections may further be classified as those:

- With negative feedback
- With positive feedback

A feedback is a path that exists between the output and input of an amplifier. This path may be purposely created by the designer of a circuit, or may exist in the circuit due to various natural reasons. For example, we know that power supply positive and negative lines (or rails) are naturally existing feedback paths. However, a designer may purposefully introduce one or more connections between the input and output terminals for getting certain benefits. Figure 5.13 shows both types of feedback paths.

It can be found from the derivations given below that feedback can really affect the performance of a system in many ways. Consider Fig. 5.14, which shows an amplifier whose output is fed back to its input through a feedback network, called the B network. The input to the amplifier V_s and the fed back voltage V_f are summed at a point called the summing point. If the polarity of the fed back voltage at the summing point is negative (positive), we say that the feedback employed in the system is negative (positive) feedback. We shall prove that, in general, negative feedback stabilises the performance of an amplifier, while positive feedback develops instability in it.

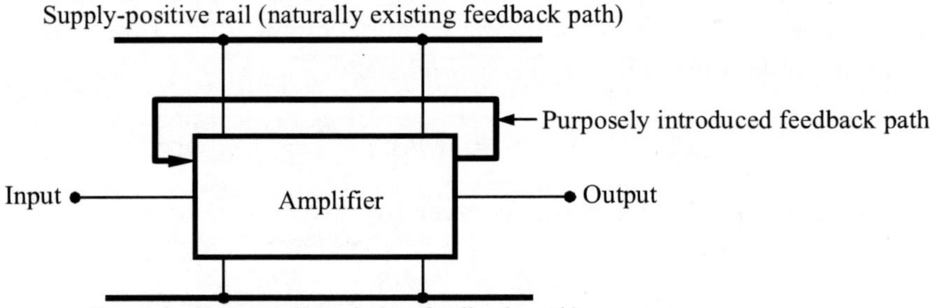

Fig. 5.13 Amplifier with feedback connections.

5.5.2 Analysis of Negative Feedback Amplifier

Consider Fig. 5.14, which shows the fed-back voltage at the summing point being negative. We can, with respect to the summing point, write the following relation:

$$V_i = V_s - V_f \tag{5.45}$$

where V_s = source (externally-applied input) voltage to the amplifier, V_i = input voltage (or error voltage) to be amplified, and V_f = voltage fed back from the output to the input. From Fig. 5.14, we also find that

$$V_f = BV_o \tag{5.46}$$

where B = feedback factor or gain of feedback (B–) network, and V_o = amplifier output voltage.

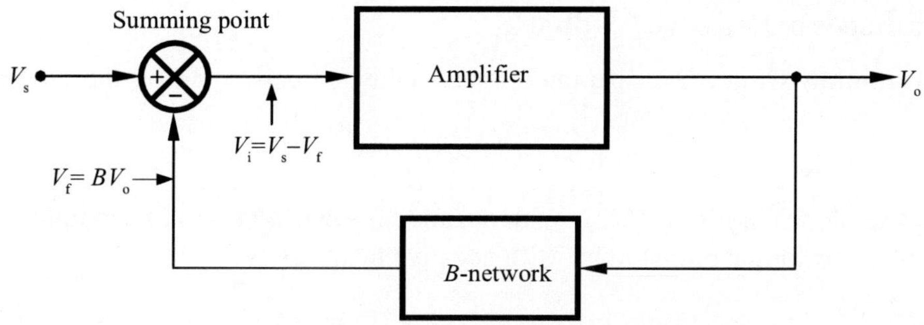

Fig. 5.14 Amplifier with negative feedback.

Combining Eqs. (5.45), and (5.46), we get

$$V_i = V_s - BV_s \tag{5.47}$$

Now, all amplifiers amplify signals that are applied across its input terminals. Therefore, we find that

$$V_o = AV_i \tag{5.48}$$

Using Eqs. (5.47), and (5.48), we find

$$V_o/A = V_s - BV_o \tag{5.49}$$

Rearranging Eq. (5.49) yields

$$V_o = AV_s - ABV_o \tag{5.50}$$

from which we get the voltage gain of the amplifier with negative feedback as

$$A_{fn} = A/(1 + AB) \tag{5.51}$$

In Eq. (5.51), let us assume that loop-gain

$$AB \gg 1 \tag{5.52}$$

Then, Eq. (5.52) reduces to

$$A_{fn} = A/AB = 1/B \tag{5.53}$$

From Eq. (5.53), we find that the gain has become *totally independent of transistor parameters, and is now entirely dependent on the feedback factor B*. This means that, if we make the B network made entirely up of fixed resistors, we can fix the gain at any desired value (of course, within reasonable limits), as well as stabilise it against temperature variations.

5.5.3 Advantages of Negative Feedback

1. **Stabilisation of Gain:** Gain parameter gets stabilised depending on the type of feedback used.

2. **Reduction in Output Noise:** *Noise is unwanted signal. It is the variation in the amplitude of a signal*, as shown in Fig. 5.15(b). With negative feedback, noise in the output gets reduced. The expression for the output noise with negative feedback is

$$N_o = N_i/(1 + AB)$$

where N_i is the noise input and N_o is the noise output with negative feedback. Since the denominator is greater than 1, we find that the noise output gets reduced.

3. **Reduction in distortion:** *Distortion is the variation from the original shape of a signal*, as shown in Fig. 5.15(d). Strictly speaking, it is the variation in the frequency of the waveform concerned.
Noise may or may not introduce additional frequency components; however, distortion will be definitely created due to the presence of extra frequency components, which are known as harmonic frequencies.

4. **Increase in Bandwidth:** With negative feedback, bandwidth of the amplifier system is seen to increase.

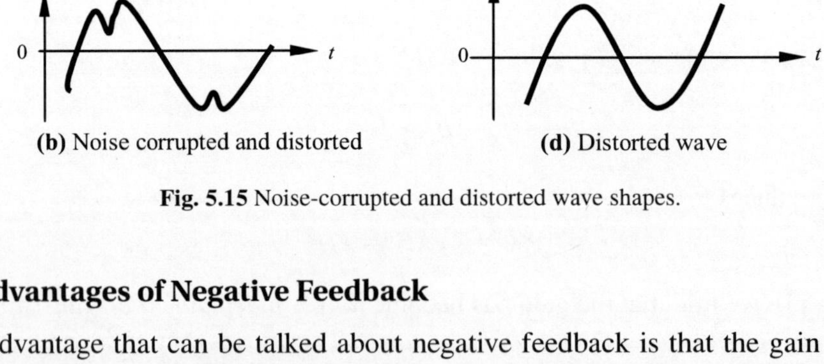

(a) Original wave

(c) Original wave

(b) Noise corrupted and distorted

(d) Distorted wave

Fig. 5.15 Noise-corrupted and distorted wave shapes.

5.5.4 Disadvantages of Negative Feedback

The only disadvantage that can be talked about negative feedback is that the gain is considerably reduced with feedback. But the reduction in gain can be compensated for by adding more number of stages. Hence this is not really a disadvantage.

5.5.5 Observation

We have proved that noise and distortion gets reduced with feedback. *However, if we cascade a number of stages to compensate for the gain, then these two factors also get correspondingly increased.*

5.6 TYPES OF NEGATIVE FEEDBACK

Negative feedback is classified into four categories:

- Voltage-series feedback
- Current-series feedback
- Current-shunt feedback
- Voltage-shunt feedback

Figure 5.16 shows the basic structure of the voltage-series feedback configuration. As can be seen from the Fig. 5.16, in voltage-series feedback, we find that a portion of the output voltage is fed back in series with the input voltage.

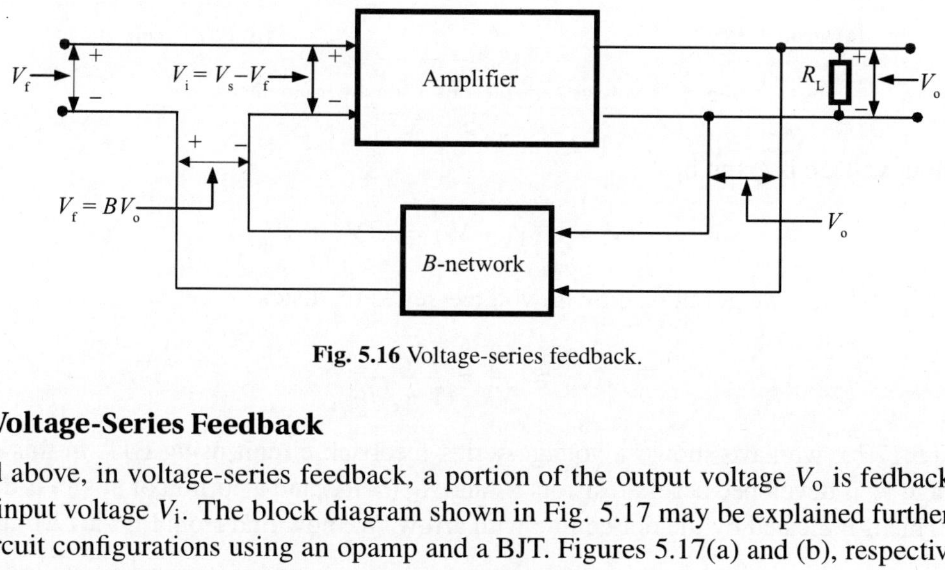

Fig. 5.16 Voltage-series feedback.

5.6.1 Voltage-Series Feedback

As stated above, in voltage-series feedback, a portion of the output voltage V_o is fedback in series with the input voltage V_i. The block diagram shown in Fig. 5.17 may be explained further by using actual circuit configurations using an opamp and a BJT. Figures 5.17(a) and (b), respectively, show the voltage-series feedback amplifier using opamp and BJT.

In Fig. 5.17(a), we find that the output voltage V_o is developed across resistors R_2 and R_1. A portion of this voltage $V_f(= BV_o)$, developed cross R_1 is fed back in series with the input voltage V_i. This is denoted by the dotted line with arrow in Fig. 5.17(a). It may be noted that V_s is the externally applied input signal. From Fig. 5.17(a), we find that

$$V_i = V_s - V_f \tag{5.54}$$

where the fed-back voltage

$$V_f = V_o R_1/(R_1 + R_2) = BV_o \tag{5.55}$$

and the feedback factor

$$B = R_1/(R_1 + R_2) \tag{5.56}$$

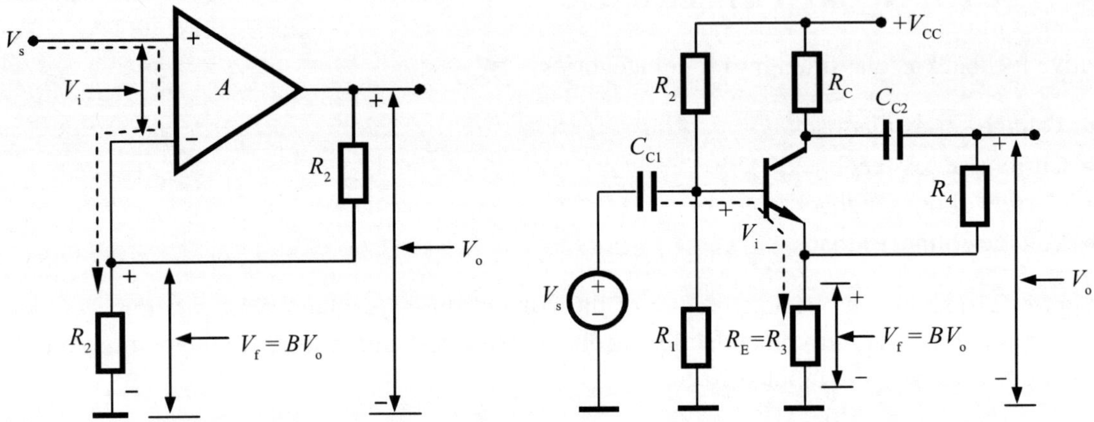

(a) Opamp circuit

(b) BJT circuit

Fig. 5.17 Voltage-series feedback amplifier circuits.

The output voltage is given by

$$V_o = AV_i = A(V_s - V_f) = A(V_s - BV_o) \tag{5.57}$$

Rearranging Eq. (5.66), we get the gain with voltage-series feedback

$$A_{Vf} = \frac{V_o}{V_s} = \frac{A}{1 + AB} \tag{5.58}$$

In Fig. 5.17(b), we have shown a voltage-series feedback circuit using BJT. In this case, the output voltage V_o is developed across resistors R_4 and R_E (= R_3) and a portion of this V_f is developed across R_3. This is indicated by the dotted line with arrow. We now find from Fig. 5.17(b) that

$$V_f = V_o R_3/(R_3 + R_4) = BV_o \tag{5.59}$$

where

$$B = R_3/(R_3 + R_4) \tag{5.60}$$

It may be noted that input voltage V_i in this case is the base-emitter voltage V_{BE}, and this comes in series with the fed-back voltage V_f, as shown by the dotted line in Fig. 5.17(b).

We also find that emitter-follower (common-collector amplifier) is a voltage-series feedback amplifier. In Fig. 5.18, which shows the circuit of an emitter-follower, we find that entire output voltage V_o (and not a portion of it) is fed back in series with the input voltage $V_i = V_{be}$. Since the entire output voltage is fed back, the gain of the amplifier is unity.

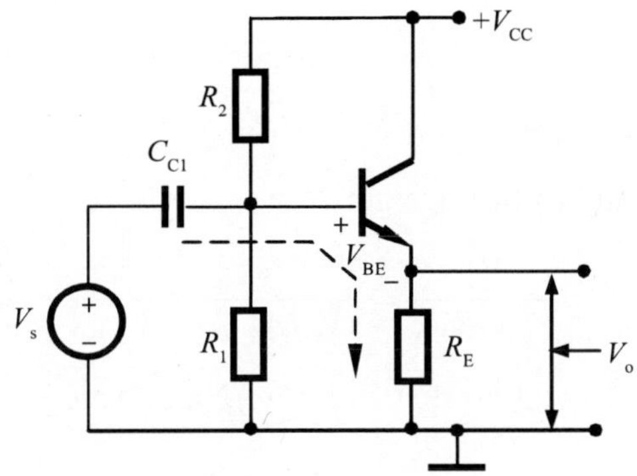

Fig. 5.18 Emitter-follower.

5.6.2 Gain of Voltage-Series Feedback Amplifier

In the voltage-series feedback circuit, as stated earlier, a portion of the output voltage is feedback in series with the input. This is illustrated in the equivalent circuit, shown in Fig. 5.19. Since, as stated earlier, the voltage-series feedback amplifier is a voltage-to-voltage amplifier, we prove that the *voltage gain gets stabilised* when negative feedback is employed in it. We define voltage gain with feedback as

$$A_{vf} = V_o/V_s \tag{5.61}$$

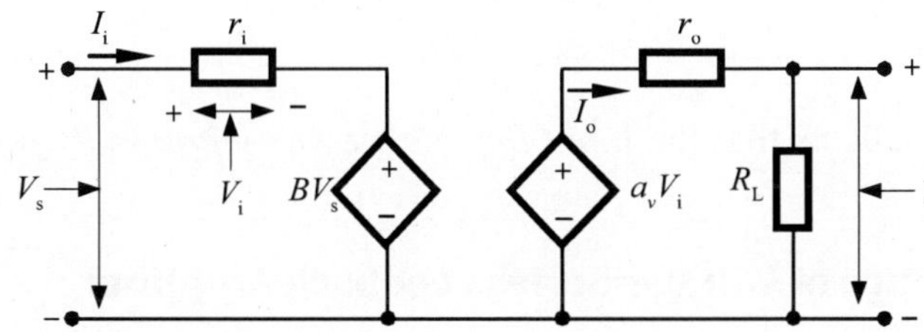

Fig. 5.19 Equivalent circuit of amplifier with voltage-series feedback.

To find this gain, we have from the input loop, using KVL

$$V_s = V_i + BV_o \tag{5.62}$$

Similarly, from the output loop, we find that output current

$$I_o = \frac{a_v V_i}{r_o + R_L} \tag{5.63}$$

Using Eq. (5.63), we find output voltage

$$V_o = I_o R_L = \frac{a_v V_i R_L}{r_o + R_L} = A_v V_i \tag{5.64}$$

where we define a modified voltage gain A_v as

$$A_v = \frac{a_v R_L}{r_o + R_L} \tag{5.65}$$

Now, using Eqs. (5.63), (5.64) and (5.65), we get

$$A_{vf} = \frac{V_o}{V_s} = \frac{V_o}{V_I + BV_o} = \frac{V_o}{(V_o/A_v) + BV_o} = \frac{A_v}{1 + A_v B} \tag{5.66}$$

From Eq. (5.66), when $A_v B \gg 1$, we find that $A_{vf} = 1/\beta$, and hence conclude that the voltage gain gets stabilised.

5.6.3 Input Impedance of Voltage-Series Feedback Amplifier

We define input impedance of the amplifier with feedback as

$$Z_{if} = V_s/I_s = V_s/I_i \tag{5.67}$$

since $I_s = I_i$. Substituting for V_s and I_i, we get

$$Z_{if} = \frac{V_i + BV_o}{I_i} = \frac{V_I + BA_v V_i}{I_i} = \frac{V_i(1 + BA_v)}{I_i} = Z_i(1 + BA_v) \tag{5.68}$$

where we have used the expression for the input impedance of the amplifier without feedback as

$$Z_i = V_i/I_i \tag{5.69}$$

Inspection of Eq. (5.69), tells us that the *input impedance has increased* with voltage-series feedback.

5.6.4 Output Impedance of Voltage-Series Feedback Amplifier

To get the expression of the output impedance, we draw the equivalent circuit shown in Fig. 5.20 based on the following rules:

1. Remove the input voltage source by *short-circuiting* the input terminals (it may be noted that if the input is a current source, to make it zero, we *open circuit* it). In either case, the internal impedance of the source must be left untouched.
2. Leave all dependent sources and circulating currents untouched.
3. Remove the load resistance from the output side, since the output impedance is defined with this condition imposed on it.

4. Apply an external voltage across the output terminals and find the resulting output current. Notice that the direction of output current I_o has become reversed (i.e., anticlockwise) in this process. This change in the direction of I_o will produce a change in the direction of any source dependent on it. Fortunately, in this case, no sources are dependent on I_o, and hence the change in its direction will not affect our analysis.
5. We define the output impedance with feedback as

$$Z_o = V_o/I_o \tag{5.70}$$

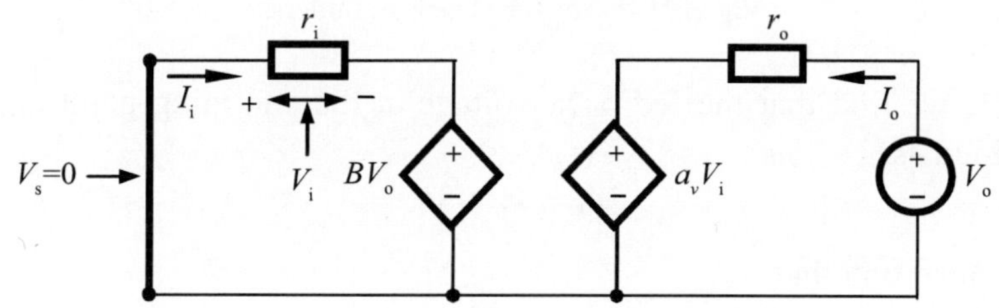

Fig. 5.20 Equivalent circuit for computing output impedance.

From the output loop, using KVL, we find

$$I_o = (V_o - a_v V_i)/r_o \tag{5.71}$$

From the input loop, we get

$$BV_o = -V_i \tag{5.72}$$

Substituting for V_i from Eq. (5.72) into Eq. (5.71), we obtain

$$I_o = \frac{V_o + a_v B V_o}{r_o} = \frac{V_o(1 + a_v B)}{r_o} \tag{5.73}$$

Rearranging Eq. (5.73), we get

$$Z_{of} = \frac{V_o}{I_o} = \frac{r_o}{1 + a_v B} \tag{5.74}$$

Inspection of Eq. (5.74) reveals that the output impedance has *decreased* with voltage-series feedback. Now, we state the following important rules related to negative feedback:

1. Input impedance will increase with voltage-series feedback.
2. Output impedance will decrease with voltage-series feedback.

5.7 POSITIVE FEEDBACK AMPLIFIER

When the fed-back voltage is positive, we call it as *positive feedback*. Figure 5.21 shows the basic block diagram of a positive-feedback amplifier.

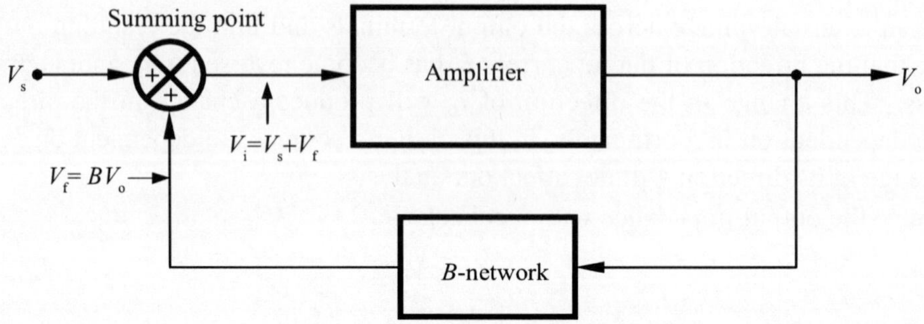

Fig. 5.21 Positive feedback amplifiers.

From Fig. 5.21, we find that the fed-back voltage at the summing point is positive, and Eq. (5.45) can be modified as

$$V_i = V_s + V_f \tag{5.75}$$

From Fig. 5.21, we also find that

$$V_f = BV_o \tag{5.76}$$

where, as before, B = feed-back factor or gain of feedback ($B-$) network, and V_o = amplifier output voltage. Then, we have

$$V_i = V_s + BV_o \tag{5.77}$$

$$V_o = AV_i \tag{5.78}$$

$$V_o/A = V_s + BV_o \tag{5.79}$$

Rearranging Eq. (5.79), and simplifying yields the voltage gain with positive feedback as

$$A_{fp} = A/(1 - AB) \tag{5.80}$$

Let us now assume that in Eq. (5.80)

$$1 - AB = 0 \tag{5.81}$$

Then

$$A_{fP} = V_o/V_s = A/0 = \infty \tag{5.82}$$

This means that the gain with feedback becomes infinity, when the total loop-gain

$$AB = +1 = 1\angle 0 \tag{5.83}$$

Equation (5.83) indicates two conditions

- *The total loop-gain AB is unity*
- *Total phase-shift around the loop is zero.*

Equation (5.83) points out that, to make the gain infinity, we must have the input V_s equal to zero. This means that the system starts producing signal by itself without having any external signal applied to its input terminals. This condition of *self-production of a signal by a system is known as oscillation*.

The two conditions or criteria given in Eq. (5.83) are known as the *Barkhausen* criteria (named after its originator) for oscillations.

5.7.1 Theory of Positive Feedback Amplifiers or Oscillators

We know that electrons are always in motion in resistors and other such devices due to the temperature existing in the atmosphere. These electronic motions can be considered as tiny noise currents, as they are random in nature. These develop tiny noise-voltage drops in resistors, which have been observed to contain frequencies from 0 Hz (DC) to about 10^{13} Hz (infrared, the frequency of the thermal radiation), and are assumed to be present in resistors existing at the input of amplifiers. The amplifier amplifies these frequencies and applies them to the input of a feedback network B, which is a frequency-selective network. It will choose only that input noise frequency to which it resonates to, and allows it (the noise frequency) to pass through it (the B network) to the input of the amplifier. In this process, the chosen noise frequency may suffer attenuation, if the B network is made up of passive devices or is an amplifier.

Now, the amplifier amplifies the fed-back frequency and applies it again to the B network where it may suffer attenuation or amplification, as stated above. But this newly attenuated output of the B network can be seen to be more amplified than its initial noise input. This output of the B network is again applied to the amplifier and the B network for still further amplification and attenuation.

Thus we find that the frequency-selected noise voltage undergoes continuous amplification by the amplifier A and attenuation (or amplification) by B, as it propagates around the loop consisting of the amplifier and attenuator (or second amplifier) until the *input and the fed-back input have the same amplitude and phase*. Then sustained oscillations result in the system, provided that the attenuation (or losses in the system) is *just counterbalanced* by the amplification (or gain) of the system. This means that if the total loop gain is unity and has zero phase-shift, the system will produce sustained harmonic or sinusoidal oscillations.

If the loop gain is far greater than unity, we find that non-sinusoidal oscillations will result in (such a set-up is possible, if B is an amplifier). On the other hand, if the loop gain is less than unity, no oscillations will build-up in the system.

The theory of oscillations using positive feedback discussed above is of a general nature and is not based on the principle of working of any particular electronic device. Hence this theory may be used to explain the principle of working of any type of oscillator using positive feedback amplifier.

5.7.2 RC Phase-shift Oscillator

We can prove that a low-pass (phase lead) and a high-pass (phase lag) RC network, shown in Figs. 5.22(a) and (b), respectively, can produce a theoretical maximum phase-shift of 90°, when $\omega = \infty$. It can also be seen that, practically, this phase-shift will be less than 90°.

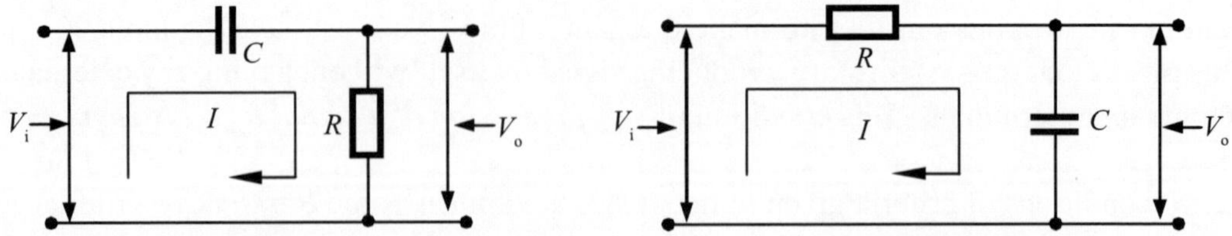

(a) Lead network (b) Lag network

Fig. 5.22 Basic *RC* lead and lag networks.

Now, let us cascade three *RC* network sections, as shown in Fig. 5.23. Since each section of these networks can produce a phase-shift of approximately 90°, by cascading three such sections, we can produce a phase-shift of more than 180°. Then by using an amplifier, which can produce another 180° phase-shift, we can convert the cascaded sections into an oscillator. Such an oscillator is called the *RC* phase-shift oscillator (PSO).

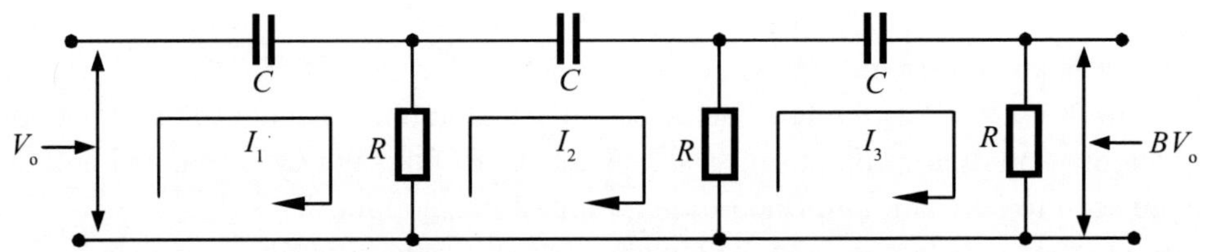

Fig. 5.23 Three-section RC phase-lead network.

It may be noted in this context that cascading of only two *RC* sections is insufficient for constructing the PSO. This is because the total phase-shift produced by cascading two *RC* sections will be *less than* 180° and therefore it cannot be attached to an amplifier that can produce a phase-shift of 180°. Thus, to construct a PSO, we must have a minimum of three *RC* sections cascaded. It may also be noted that we can construct PSO's with more than three *RC* sections, but all of such PSO's can obviously be seen to be uneconomical compared to the three-section PSO.

PSO using Phase-Lead Network

Figure 5.24 shows an *RC* PSO constructed by using a BJT amplifier, and a three-section *RC* phase-lead network. This network is called a phase-lead network, because of the phase-leading nature of its input compared to its output. A lag network can also be used for the construction of a PSO.

As shown in Fig. 5.24, the PSO consists of a three-section phase-lead network connected to a BJT *CE* amplifier. The amplifier is designed such that its *Q* point is in the middle of its active region. For fixing the gain of the amplifier, a potentiometer is connected across emitter resistor R_E.

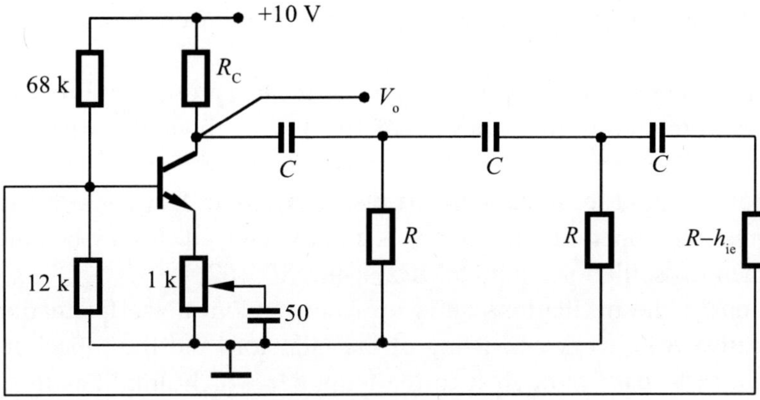

Fig. 5.24 *RC* phase-shift oscillator using lead network.

From analysis, we can prove that for oscillations to take place,

- The gain of the amplifier must be greater than 29.
- The frequency of oscillation is given by $f_o = 1/2\pi RC\sqrt{6+4(R_C/R)}$

Hartley Oscillator

Generally, *RC* oscillators are designed to operate in the audio-frequency range (below 1-MHz) and *LC* oscillators are designed to operate in the radio-frequency range (above 1-MHz). However, *LC* oscillators can also be designed to produce oscillations in the low audio-frequency range. But, for the low-frequency operation, the inductors used become very large in value and hence in physical size. This puts a limit on the lowest operating frequency of oscillators employing *LC*-coupling networks.

5.7.3 Hartley Oscillator–Common-Emitter Configuration

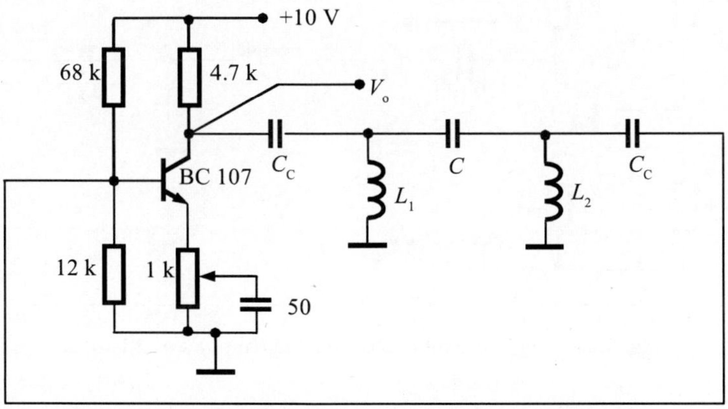

Fig. 5.25 Common-emitter *RC*-coupled Hartley oscillator.

Figure 5.25 shows the structure of a Hartley oscillator in the *CE* configuration. It shows the *RC*-coupled type.

The transistor used must have very good high frequency properties From data manuals, we find that the unity-gain transition frequency f_T of BC 107 is about 150 MHz. According to the Barkhausen criteria, the gain required for oscillation is only unity. This means that we can generate oscillations up to 150 MHz, at least theoretically, using BC 107. However, due to loading effects, it is seen that the maximum operating frequency is greatly reduced. It may be noted in this context that it is easy to generate oscillations up to 5 MHz, using BC 107.

As stated previously, the oscillations build up from random noise fluctuations in the circuit. The frequency selective *B* network consisting of the inductors and the capacitor will select their frequency of resonance to pass through it to the amplifier which amplifies this frequency alone and applies it again to the *B* network which attenuates it. The amplification and attenuation in the equilibrium state will generate sustained oscillations from the circuit. From analysis, we can prove that for oscillations to take place,

- The gain of the amplifier must be greater 1 as the oscillator uses lossless components of *L* and *C* for *B* network.
- The frequency of oscillation is given by $f_o = 1/2\pi\sqrt{LC}$.

5.7.4 Colpitt's Oscillator

The Colpitt's oscillator is similar to the Hartley oscillator in construction, with the inductors being replaced with capacitors and the capacitor with an inductor. Figure 5.26 shows the CE RC-coupled variety. The frequency of oscillation is given by $f_o = 1/2\pi\sqrt{LC}$ where $C = C_1C_2/(C_1 + C_2)$. Usually, since Colpitt's oscillator is a high-frequency oscillator, the capacitors will be in the pico-farads range. The condition for gain is $A_v = C_1/C_2$.

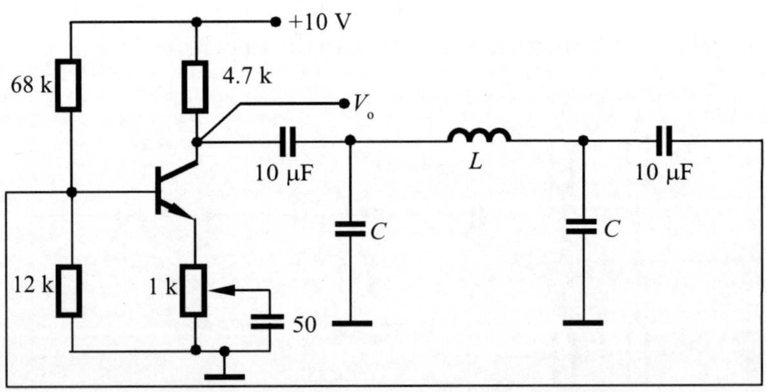

Fig. 5.26 Common-emitter RC-coupled Colpitt's oscillator.

We see that the gain can be adjusted by suitably choosing the values of C_1 and C_2. In our initial design with *CE* amplifier, we choose $C_1 = C_2$. As in the case of the Hartley, this criterion allows us a lot of freedom in the design of the amplifier.

5.7.5 Crystal Oscillators

Crystal oscillators employ piezo (to be pronounced as pi-ee-zo)-electric crystals to produce highly stable oscillations. Figure 5.27 shows the equivalent circuit of the crystal. The crystal may be represented by a series-resonant *RLC* circuit shunted by a capacitor C_P, as shown. L of the resonant circuit represent the mass of the crystal material, C its elastic property, R the losses in the crystal under operating conditions, and C_P the capacitance between the crystal metal plates. Also shown in the figure are inductance L_S and resistance R_S. These represent, respectively, the lead inductance and lead resistance and are shown to be in series with the main equivalent circuit.

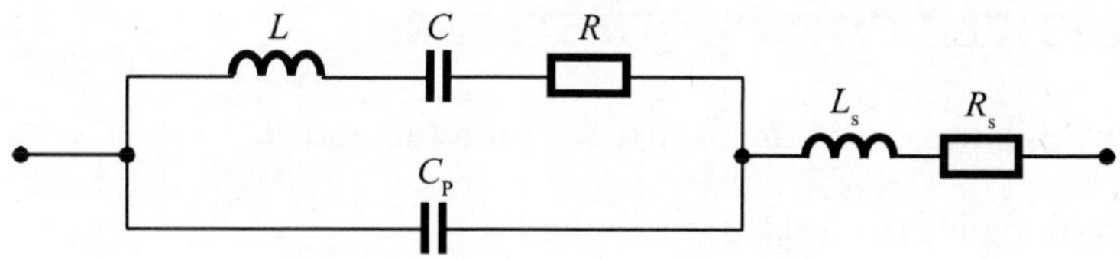

Fig. 5.27 Equivalent circuit of a quartz crystal.

For a typical crystal, L will be in henrys, C in tens of femto farads, C_P in pico farads, and R in kilo ohms. L_s and R_s are normally negligible, unless the resonant frequency is extremely high. Since L is very high, and C is very small, the quality factor Q of the crystal is very high. For a 100-kHz crystal, with $L = 8$ H, $C = 50$ femto farad, and $R = 1000\ \Omega$, the Q-factor will be above 5000. It is this high value of Q that enables a crystal to maintain a high degree of stability in oscillations.

A crystal oscillator has *two resonant frequencies*. The first one is the series-resonant frequency of the crystal itself, given by $f_{os} = 1/2\pi\sqrt{LC}$. The second one is the parallel-resonant frequency, given by $f_{op} = 1/2\pi\sqrt{LC'}$ where $C' = C \parallel C_P$. Usually, $f_{op} > f_{os}$ and the actual frequency of oscillation f_o lies in between these two extreme values.

Colpitt's Crystal Oscillator

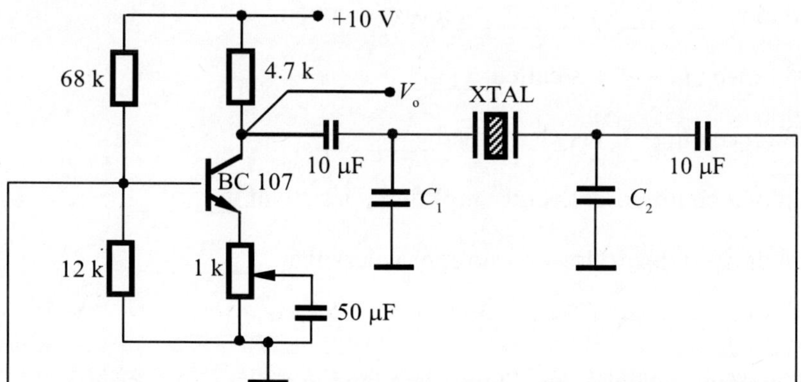

Fig. 5.28 Crystal oscillator (Colpitt's type).

This circuit is shown in Fig. 5.28. This is similar to the *LC* Colpitt's oscillator with the inductor in the *B*-network replaced by the crystal, and hence the name. In this oscillator, the crystal behaves primarily as inductor and produce oscillations in association with capacitors C_1, C_2, and C_P. However, we still find that the frequency of oscillation is approximate equal to that of the resonant frequency of the crystal itself. Hence the choice of C_1 and C_2 are not entirely critical; yet, from experience, they are chosen to be in the pico farads range. Typically, we may choose $C_1 = C_2 = 100$ pF. Finally, we add a pot in the feedback path as shown. By adjusting the pot, excess gain may be reduced to the required level to produce pure sine waves.

ONE-WORD/MULTIPLE-CHOICE QUESTIONS

M5.1 The emitter resistor is used in a transistor for temperature ________.

M5.2 The voltage-divider bias is used for fixing the ________.

M5.3 The collector resistance is used for fixing the Q-point in the ________.

M5.4 Coupling capacitor is used for ________ and passing ________.

M5.5 Thermal voltage V_T may be computed by using the expression ________.

M5.6 The product $g_m R_L$ is used to compute the ________ of an amplifier.

M5.7 The product βf_β is known as ________.

M5.8 The product $h_{fe} f_H$ is known as ________.

M5.9 The product $h_{fe} f_H$ is known as ________ frequency.

M5.10 The frequency at which current gain becomes $1/\sqrt{2} \times$ the current gain at mid-frequency is called the ________ or ________ cut-off frequency.

M5.11 The difference $(f_H - f_L)$ is called ________.

M5.12 The difference $(f_H - f_L)$ is also called ________.

M5.13 The gain of a common-collector amplifier is less than ________.

M5.14 The gain of a emitter-follower amplifier is less than ________.

M5.15 The gain of a common-drain amplifier is less than ________.

M5.16 The criteria for producing oscillations are known as ________ criteria.

M5.17 The Barkhausen criteria for producing oscillations are ________.

M5.18 The Barkhausen criteria for producing oscillations are ________.

M5.19 The Barkhausen criteria for producing oscillations are

1. Loop gain is ________.
2. Total phase shift around the loop is ________ or ________.

M5.20 The frequency of oscillation of a Wien-bridge oscillator is ________.

M5.21 The frequency of oscillation of a phase-shift oscillator using lead network is ________.

M5.22 The frequency of oscillation of a phase-shift oscillator using lag network is ________.

M5.23 The gain required to produce oscillation in a phase-shift oscillator using lag network is ________.

M5.24 The gain required to produce oscillation in a phase-shift oscillator using lead network is ________.

M5.25 The gain required to produce oscillation in a Wien-bridge oscillator using lag network is ________.

M5.26 A quartz crystal may be represented by an equivalent circuit consisting of a series ________ circuit.

REVIEW QUESTIONS

1. Explain how a bipolar junction transistor can act as an amplifier.
2. What do you mean by operating point?
3. What is load line?
4. What is meant by biasing of amplifier? Why do you require it?
5. Explain fixed bias.
6. Explain biasing.
7. Explain diode biasing.
8. What is current-mirror biasing?
9. Discuss different types of biasing techniques used in the case of BJT amplifiers.
10. What is stability of an amplifier?

11. What are stability factors?
12. Derive the current-stability factor in the cases of fixed and R_1/R_2 bias with R_E incorporated.
13. Derive voltage-stability factor in the case R_1/R_2 bias.
14. Derive current-gain stability factor in the case R_1/R_2 bias.
15. Explain thermal runaway.
16. Analyse the current-mirror circuit.
17. Explain dual-supply biasing.
18. Explain the methods for biasing JFET's and MOSFET's.
19. What do you mean by self bias?
20. What are the different biasing methods for MOSFET's.
21. What are h-parameters?
22. Derive the small-signal hybrid equivalent circuit of the bipolar junction transistor.
23. Demonstrate how h-parameters can be computed form VI characteristics.
24. Explain the frequency response of BJT.
25. Derive the hybrid-π equivalent circuit of the BJT.
26. Bring-out the equivalence between the hybrid and hybrid-π? parameters.
27. Compare BJT and MOSFET?
28. Give a few applications of BJT.
29. Explain why RC-coupling is used in amplifiers?
30. What is the advantage in using RC-coupling and what is its disadvantage?
31. Explain the need of each of the circuit elements in a class-A RC-coupled BJT voltage amplifier.
32. Analyse a BJT RC-coupled amplifier in the CE configuration using hybrid-π parameters.
33. Analyse the RC-coupled amplifier in its mid-frequency region of operation.
34. Obtain the expression for the voltage gain, input impedance, output impedance, current gain, and power gain in the case of an RC-coupled voltage amplifier using hybrid-π parameters.

35. Perform the high-frequency analysis of an *RC*-coupled BJT amplifier.

36. What is the gain-band width product of an *RC*-coupled BJT amplifier? Explain its significance.

37. What is f_T (transition frequency)?

38. Obtain an expression for the unity-gain bandwidth.

39. Explain and deduce the voltage gain analysis?

40. Perform the low-frequency analysis of an *RC*-coupled BJT amplifier?

41. Analyse the *CC* (common-collector) amplifier using hybrid parameters.

42. Analyse the *CC* (common-collector) amplifier using hybrid-π parameters.

43. Analyse the *CB* (common-base) amplifier using hybrid-π parameters.

44. Explain the phase reversal of output signal in *CE* amplifier?

45. Show why there are no phase reversals in the *CB* and *CC* amplifiers.

46. Explain the operation of a JFET common-source amplifier.

47. Perform the small-signal analysis of the JFET in the *CS* configuration.

48. What is meant by source follower?

49. Perform the small-signal analysis of the JFET in the source-follower configuration.

50. Briefly explain the theory of positive feedback in amplifiers?

51. Explain the theory of electronic oscillators?

52. Explain how oscillations build up in electronic circuits.

53. What are the Barkhausen criteria for oscillations?

54. Derive the expression for the Barkhausen criteria.

55. Discuss how oscillations build-up in the Wien-bridge oscillator.

56. Draw the circuit diagram of the Wien-bridge oscillator using a simplified opamp configuration.

57. Derive the expression for the frequency of oscillation of a WBO.

58. Analyse the WBO using appropriate equivalent circuit.

59. Derive the expression for the gain required to build up and sustain oscillations in a WBO.

60. Draw the circuit of an *RC* phase-shift oscillator using BJT, and explain how oscillations are generated in it.

61. What is meant by a phase-lead network?

62. Explain a three-section *RC* phase-lead network.

63. Analyse the *RC* phase-shift oscillator. Derive the expression for the frequency of oscillation and obtain the gain condition.

64. Explain the theory of *LC* oscillators.

65. Explain the theory of Hartley oscillator.

66. Explain the theory of Colpitt's oscillator.

67. Explain the theory crystal oscillators.

68. Draw the equivalent circuit a quartz crystal and explain how it is derived.

69. Explain the advantages and disadvantages of *RC*, *LC*, and crystal oscillators.

UNSOLVED PROBLEMS

P5.1 Compute the ratio of channel lengths, if for width $W_1 = 1$ μm, $I_{D1} = 2$ mA, and for width $W_2 = 2$ μm, drain current $I_{D2} = 1$ mA.

P5.2 Compute the ratio of drain currents, if for channel width $W_1 = 1$ μm, channel length $L_1 = 2$ μm and for width $W_2 = 2$ μm, $L_2 = 1$ μm.

P5.3 Compute the value of transconductance in the case of a BJ transistor if the collector current is 1 mA at normal room temperature.

P5.4 Compute the value of β in the case of a BJT if the collector current is 1 mA at normal room temperature and input resistance r_π is 3 kΩ.

P5.5 Compute the value of β in the case of a BJT if the collector current is 1 mA at normal room temperature and input resistance r_π is 1 kΩ.

P5.6 Compute the value of h_{re} in the case of a BJT if the collector current is 1 mA at normal room temperature and input resistance r_π is 3 kΩ and $r_\mu = 1000$ kΩ.

P5.7 Compute the value of output admittance Y_o in the case of a BJT, if collector current is 1 mA at normal room temperature, r_π is 3 kΩ, $r_\mu = 1000$ kΩ, and $r_o = 1000$ kΩ.

P5.8 Find the frequency of oscillation of a Wien-bridge oscillator, if the feedback network consists of $R = 1$ kΩ and $C = 0.01$ μF.

P5.9 Find the frequency of oscillation of a Wien-bridge oscillator, if the feedback network consists of $R = 3.3$ kΩ and $C = 0.005$ μF.

P5.10 Find the frequency of oscillation of an RC phase-shift oscillator, if the feedback network consists of three section, each section comprising of $R = 2.7$ kΩ and $C = 0.001$ μF.

P5.11 Find the frequency of oscillation of a Hartley oscillator, if the feedback network consists of $L_1 = 1$ μH, $L_2 = 3$ μH, and $C = 100$ pF. Specify the gain of the amplifier to be used.

P5.12 Find the frequency of oscillation of a Hartley oscillator, if the feedback network consists of $L_1 = 1$ μH, $L_2 = 3$ μH, $M = 5$ μH and $C = 100$ pF.

P5.13 Find the frequency of oscillation of a Hartley oscillator, if the feedback network consists of $L_1 = 3$ μH, $L_2 = 5$ μH, $M = -3$ μH and $C = 100$ pF.

P5.14 Find the frequency of oscillation of a Colpitt's oscillator, if the feedback network consists of $C_1 = 100$ pF, $C_2 = 100$ pF, $L = 5$ μH. Specify the gain of the amplifier used.

P5.15 Find the frequency of oscillation of a Colpitt's oscillator, if the feedback network consists of $C_1 = 400$ pF, $C_2 = 100$ pF, $L = 5$ μH. Specify the gain of the amplifier used.

Introduction to Operational Amplifiers

CHAPTER 6

6.1 IDEAL OPERATIONAL AMPLIFIER

Operational amplifiers (opamps) are high-gain direct-coupled (DC) amplifiers that are used to perform mathematical operations such as addition, subtraction, integration, differentiation, etc. The *ideal opamp* should possess the following characteristics:

(a) Voltage gain = ∞
(b) Input impedance = ∞
(c) Output impedance = 0
(d) Frequency response = 0 to ∞
(e) Bandwidth = ∞
(f) Drift = 0

Figure 6.1 shows the symbol of an opamp. The symbol shows two input terminals and one output terminal. The input terminals are, respectively, called the **inverting** terminal and the **non-inverting** terminal. The signal applied to the inverting terminal will appear phase-inverted and the signal applied to the non-inverting terminal will appear without phase-inversion at the output, as shown. In this text, the voltage applied at the inverting input is designated as V^- and that applied at the non-inverting input as V^+.

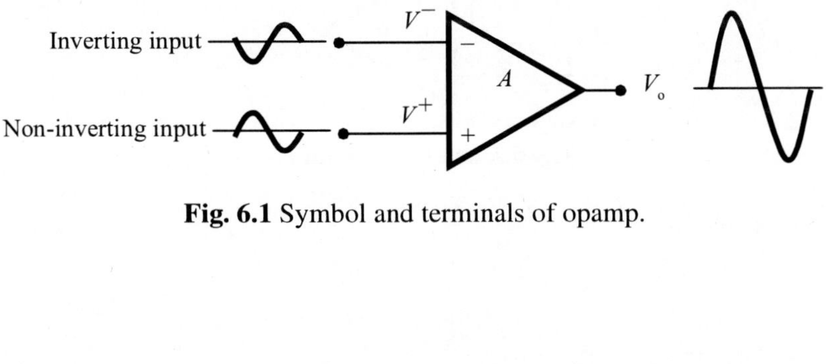

Fig. 6.1 Symbol and terminals of opamp.

6.2 VIRTUAL SHORT CIRCUIT AND VIRTUAL GROUND

Consider an ideal opamp with infinite gain. Let the input voltage be V_i and the output voltage V_o, as shown in Fig. 6.2. We know that the gain of an opamp is given by

$$A = V_o/V_i \tag{6.1}$$

Since $A = \infty$ for the ideal opamp, we find that

$$V_i = V_o/\infty = 0 \tag{6.2}$$

This result indicates that the input voltage is zero. A voltage can be zero only under a short-circuit condition. However, in the case of opamps, there exists no real short circuit across its input terminals. We therefore conclude that the input terminals of the opamp are virtually shorted together. In turn, we state that there exists a **virtual short circuit** across the input terminals of an opamp. If one of the input terminals is grounded, the **virtual short** becomes a virtual ground.

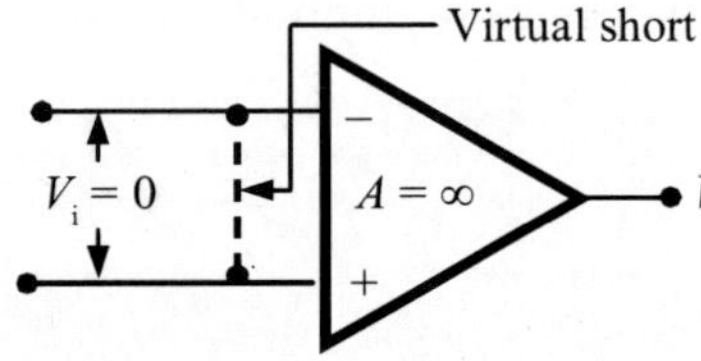

Fig. 6.2 Virtual short circuit across input terminals.

6.3 INVERTING AMPLIFIER

Figure 6.3 shows the *voltage-shunt feedback* configuration of the opamp. When a signal voltage V_S is applied to the inverting-input terminal an input current I_1 starts flowing into the opamp. However, since the input impedance of the opamp is infinity, this current will not flow into the amplifier; rather it flows through the output loop (through R_2) to the output terminal of the opamp, as shown. This means that the input current I_1 = the output current I_2.

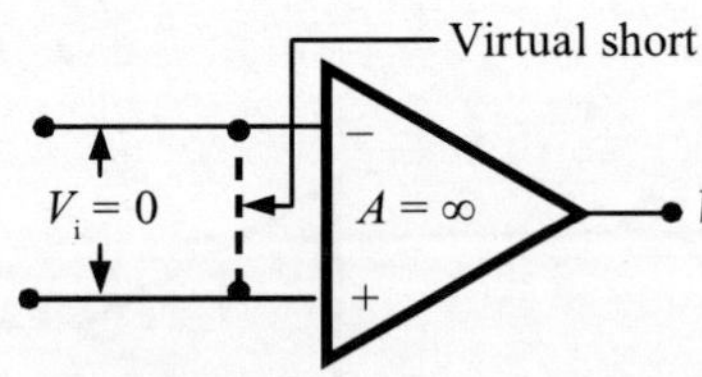

Fig. 6.3 Inverting amplifier.

With respect to node X, we find

$$I_1 = (V_S - V_X)/R_1 \tag{6.3}$$

and

$$I_2 = (V_X - V_o)/R_2 \tag{6.4}$$

Equating the two, we get

$$(V_S - V_X)/R_1 = (V_X - V_o)/R_2 \tag{6.5}$$

But the voltage at node X (due to the virtual ground) $V_X = 0$. Substituting this in Eq. (6.5), we find

$$V_S/R_1 = -V_o/R_2 \tag{6.6}$$

which yields the voltage gain of the inverting amplifier as

$$A_v = V_o/V_S = -R_2/R_1 \tag{6.7}$$

The negative sign in Eq. (6.7) indicates that the output is 180° out-of-phase with the input.

6.4 NON-INVERTING AMPLIFIER

Figure 6.4 shows the voltage-series feedback configuration of the opamp. When V_S is applied to the non-inverting-input terminal, I_1 flows into the opamp and I_2 flows out of the opamp, as in the previous case. Since $I_1 = I_2$, with respect to node X, because of the virtual short circuit, we find that $V_X = V_S$. Then

$$I_1 = V_S/R_1 \tag{6.8}$$

and

$$I_2 = (V_X - V_o)/R_2 \tag{6.9}$$

From Eqs. (6.8) and (6.9), we get

$$V_S/R_1 = (V_S - V_o)/R_2 \tag{6.10}$$

Rearranging and solving yields the voltage gain of the non-inverting amplifier as

$$A_v = 1 + (R_2/R_1) \tag{6.11}$$

We observe that the output is in phase with the input.

Notice that the virtual short circuit in this case is not virtual ground, as none of the input terminals are grounded. Since this configuration is a *voltage-series feedback connection*, it has high input and low output impedances, and is regarded as a voltage amplifier.

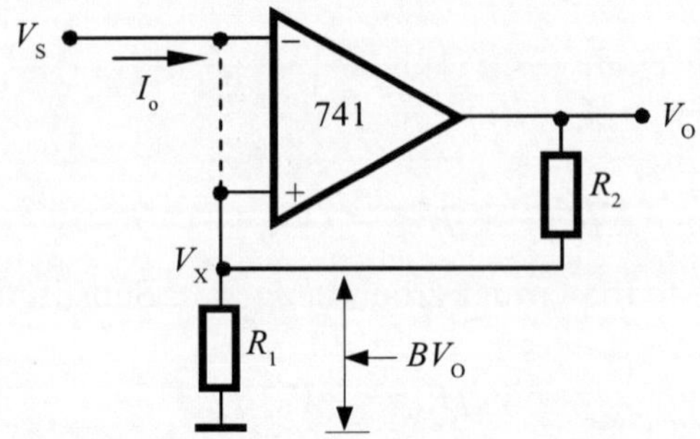

Fig. 6.4 Non-inverting amplifier.

6.5 ANALOG INVERTER

Suppose we choose resistors $R_1 = R_2 = R$ in Fig. 6.3. This results in the configuration shown in Fig. 6.5 and the gain becomes

$$A_V = -R/R = -1 \tag{6.12}$$

Equation (6.12) states that $V_o = -V_S$, which means that the output is an inverted version of the input. Thus this amplifier acts as an analog inverter.

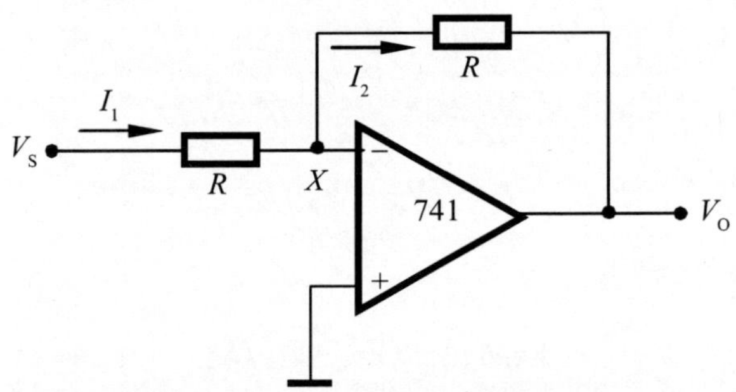

Fig. 6.5 Inverting amplifier.

6.6 ANALOG SUMMER (ADDER)

Consider Fig. 6.6. This is a modification of the inverting amplifier. Using node-voltage analysis, with respect to node X we find

$$(V_1/R_1) + (V_2/R_2) = -V_o/R_F \tag{6.13}$$

Rearranging, we obtain

$$(V_1R_F/R_1) + (V_2R_F/R_2) = -V_o \tag{6.14}$$

If $R_1 = R_2 = R_F = R$, then

$$V_1 + V_2 = -V_o \tag{6.15}$$

Thus this configuration results in the addition of the two input voltages.

The above result may be extended to yield the expression for the addition of n input voltages. Thus, we obtain the sum of n input voltages as

$$V_1 + V_2 + \cdots + V_n = -V_o \tag{6.16}$$

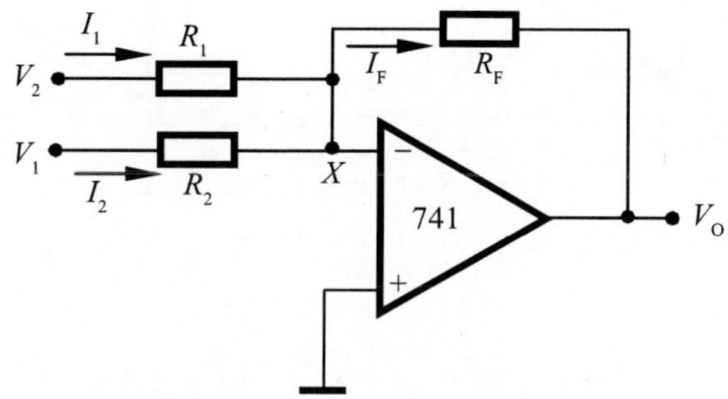

Fig. 6.6 Summer.

6.7 VOLTAGE-FOLLOWER

Suppose we remove the resistors R_1 and R_2 of Fig. 6.5, and short the inverting terminal with the output terminal, as shown in Fig. 6.7. Then, due to the virtual short circuit at the input, we find that $V_o = V_S$. Thus the output is equal to the input, which means that the output is following the input, and hence the name voltage-follower. Voltage-follower is a unity-gain amplifier, which has very large input impedance and very low output impedance. It may be noted that voltage-follower is the opamp-version of emitter-follower.

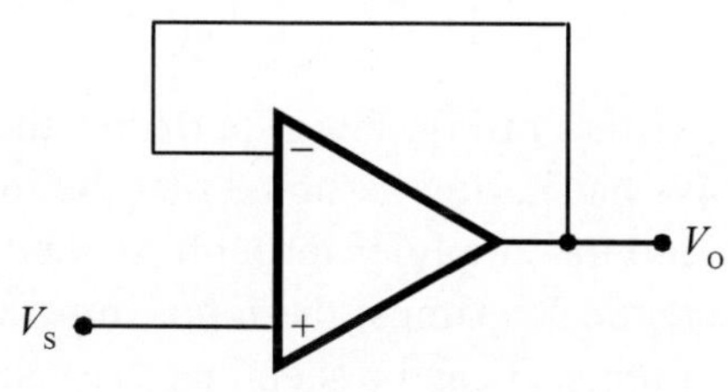

Fig. 6.7 Voltage-follower.

6.8 SUBTRACTOR

We can construct an analog subtractor by combining an inverter and a summer, as shown in Fig. 6.8. Opamp A_1 is an inverter and its output $V_{o1} = -V_1$ when its input is V_1. Now this is applied as an input to opamp A_2, which acts as a summer. The final output of the circuit, then, is

$$-V_{o2} = V_2 - V_1 \tag{6.17}$$

Equation (6.17) illustrates that the circuit shown in Fig. 6.8 behaves as a subtractor.

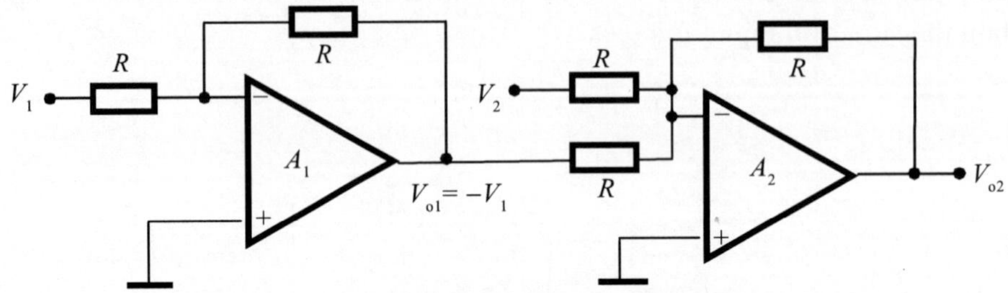

Fig. 6.8 The subtractor.

6.9 INTEGRATOR

Figure 6.9(a) shows an analog integrator. With respect to the input node,

$$I_R = V_S/R \tag{6.18}$$

and

$$I_C = C\,(dV_S/dt\,) \tag{6.19}$$

Combining the two equations yields

$$V_S/R = -C\,(dV_o/dt\,) \tag{6.20}$$

Rearranging, we obtain

$$V_o = -(1/RC\,)\int V_S dt \tag{6.21}$$

Equation (6.21) shows that the circuit of Fig. 6.9(a) performs the operation of integration. This circuit, however, has one defect. We had earlier assumed that the input impedance of an opamp is infinity and that no current flows into the amplifier through its input terminals. But, in a practical opamp, this is not true. In most practical opamps, the input impedance is finite and therefore the input transistors draw small bias currents. It can be seen that these small bias currents (I_{charge} in the figure) charge capacitor C, as shown in Fig. 6.9(b). Once C gets fully charged, the output becomes saturated and remains in its maximum value, viz., $\pm V_{sat}$, depending on the polarity of charging. It may be noted that $\pm V_{sat}$ is the maximum output voltage of an opamp, and is only slightly less than the supply voltage $\pm V_{CC}$.

To prevent C from saturating, a large resistance R_H of 1 MΩ is connected across C as shown in Fig. 6.9(b). This resistance will provide a discharge path for the charges in C and prevent the circuit from going into permanent saturation. Since the R_H is very high, it will not affect the performance of the circuit as an integrator.

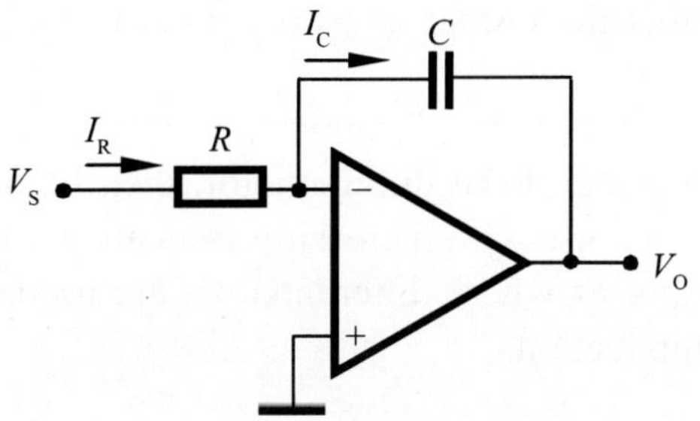

Fig. 6.9(a) Analog integrator.

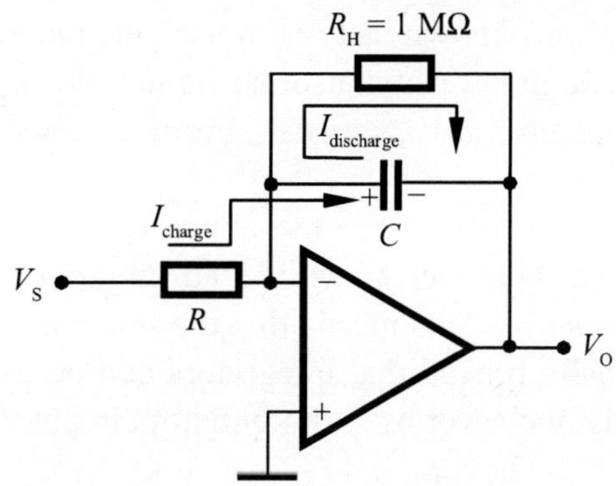

Fig. 6.9(b) Practical analog integrator

6.10 DIFFERENTIATOR

Figure 6.10 shows the circuit of a differentiator using opamp. With respect to the input node

$$I_R = -V_S/R \tag{6.22}$$

and

$$I_C = C\,(dV_C/dt) \tag{6.23}$$

Combining the two equations yields

$$V_o/R = -C\,(dV_o/dt) \tag{6.24}$$

Rearranging, we obtain

$$V_o = RC\,(dV_S/dt) \tag{6.25}$$

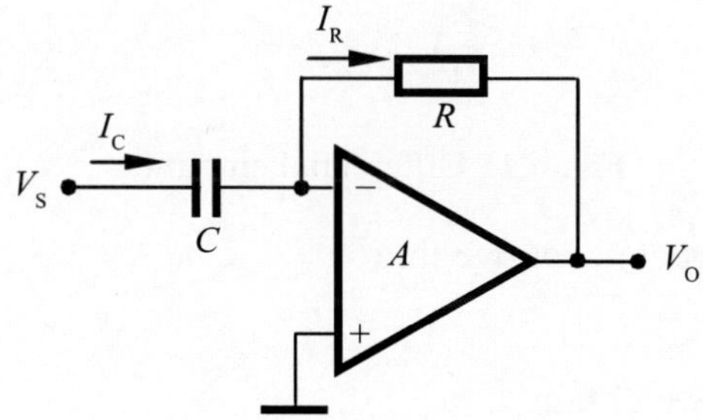

Fig. 6.10 Analog differentiator.

Note: Even though we have proved that the circuit shown in Fig. 6.10 can perform differentiation, it is not practically used. This due to the following reason:

It is a well-known fact that noise gets increased with increasing frequency. Consider the output of differentiator for a sinusoidal input. We know that the output of a differentiator for an input of $V_{max} \sin \omega t$ is $\omega V_{max} \sin \omega t$. From this, we find that the output is proportional to the angular frequency ω.

As ω increases, from the above relation, we find that the differentiator produces more and more noise, However, if we use an integrator in the place of the differentiator, then its output is $(V_{max} \sin \omega t)/\omega$. This means that the integrator reduces noise with increasing frequency. Further it can be easily proved that integrators can be used in places where differentiators are used. Hence practically, we never use differentiators in practical applications.

6.11 DIFFERENTIAL AMPLIFIER

Even though the opamp is made up of differential amplifiers, we can get a much better differential amplifier using opamps themselves. Figure 6.11 shows an opamp differential amplifier. From the figure, we observe that the nodes *X* and *Y* are virtually short circuited. Now, with respect to node *X*, we find

$$V_x = V_2 R_2/(R_1 + R_2) \tag{6.26}$$

and with respect to node *Y*, we find

$$\left(V_1 - V_y\right)/R_1 = \left(V_y - V_O\right)/R_2 \tag{6.27}$$

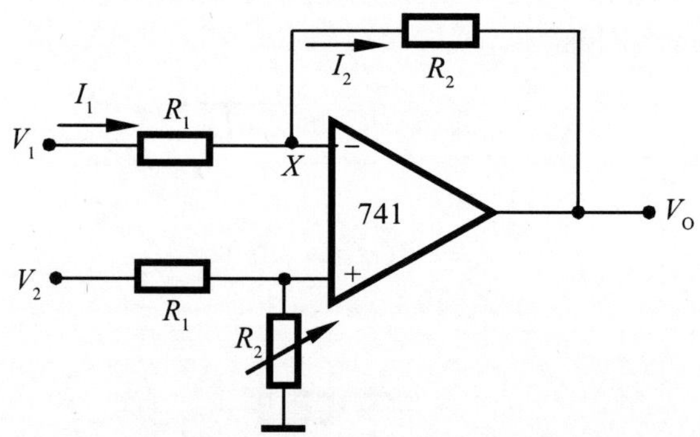

Fig. 6.11 Differential amplifier.

However, using the virtual-short concept, we find

$$V_x = V_y \tag{6.28}$$

Then simplifying and solving, we obtain

$$V_o = (V_2 - V_1)(R_2/R_1) \tag{6.29}$$

This proves that the output is the amplified version of the difference between the input voltages; hence the name differential amplifier. This configuration is widely used in modern instrumentation amplifiers.

6.12 CATHODE RAY OSCILLOSCOPE

6.12.1 Basic Principles of Operation

The block schematic of a CRO is shown in Fig. 6.12. The CRO consists of a cathode-ray tube and associated circuitry. The signal to be observed is applied to the *Y* input of the CRO. This input is amplified and attenuated to a level suitable for display on the CRT screen using the *Y-Y* amplifier-attenuator unit. The output of the *Y* amplifier is used to drive the vertical deflection (*Y-Y*) plates of the cathode-ray tube (CRT).

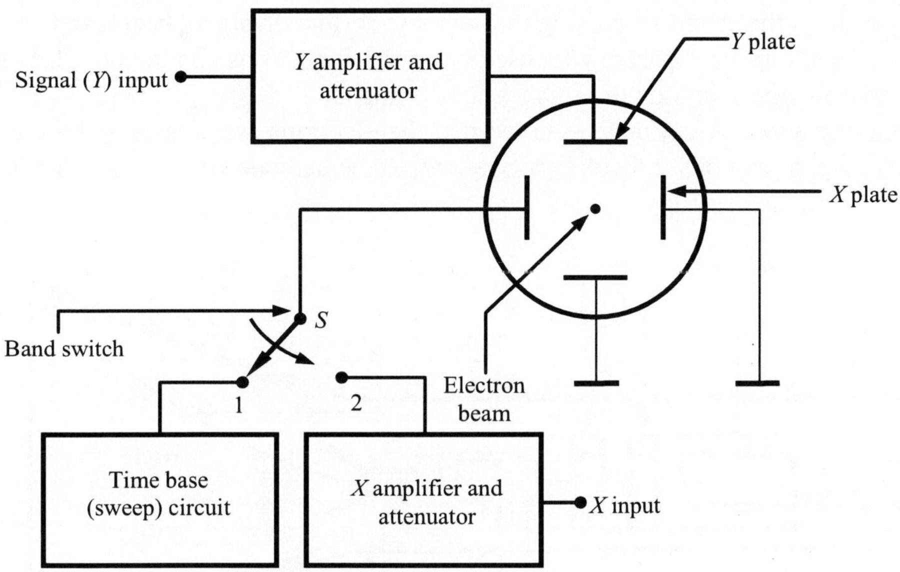

Fig. 6.12 Block diagram of a CRO.

The time-base (sweep) voltage is applied to the horizontal deflection *(X-X)* plates of the CRT through a bandswitch *S*, as shown. If we are interested in the interaction between two waveforms, then we apply the first waveform first waveform to the *Y* input and the second waveform to the *X* input of the CRO. To apply the waveform to the *X* input we have to cut the time-base voltage and connect *S* to position 2. The electron beam in the CR tube will then move in the direction of resultant of the waves applied to the *X*- and *Y*- inputs terminals.

6.12.2 Structure of the CRT

Figure 6.13 shows a simplified structure of a cathode-ray tube (CRT). A typical CRT consists of an electron gun, two pairs of deflection plates, and a phosphor screen.

Electron Gun

The electron gun is a unit that consists of a heater, an oxide-coated cathode, a control grid, and associated anodes. The cathode will emit electrons upon getting heated by the heater. These electrons are formed into a narrow beam by the focusing anode. This beam is then accelerated on to the screen by the accelerating anodes.

The focusing electrode is a group of three hollow co-axial cylinders through which the electron beam can travel to the anode.

As shown in Fig. 6.13, anodes A_1 and A_3 are kept at +2 kilovolts with respect to the cathode. The middle anode A_2 is kept at cathode (ground) potential. On the anode surfaces we find a distribution of equipotential lines, which are imaginary lines of voltages exhibiting the same value of potential and distributed uniformly over a metallic surface when a voltage is applied to it. Each of these lines deflects the electron beam by its electrostatic nature in a direction perpendicular to it. It can be seen that anode A_1 deflects the electrons passing through it to form a divergent beam, and anodes A_1 and A_3 converge the electrons into a narrow focused beam. The electrons coming out of A_3 thus can be found to be a narrow beam passing through the axis of the CRT.

Before entering anode A_1, it may be noted that the electrons are controlled by a control grid, which is kept at a potential of −2.5 volts with respect to the cathode. By varying this potential, we can control the beam intensity.

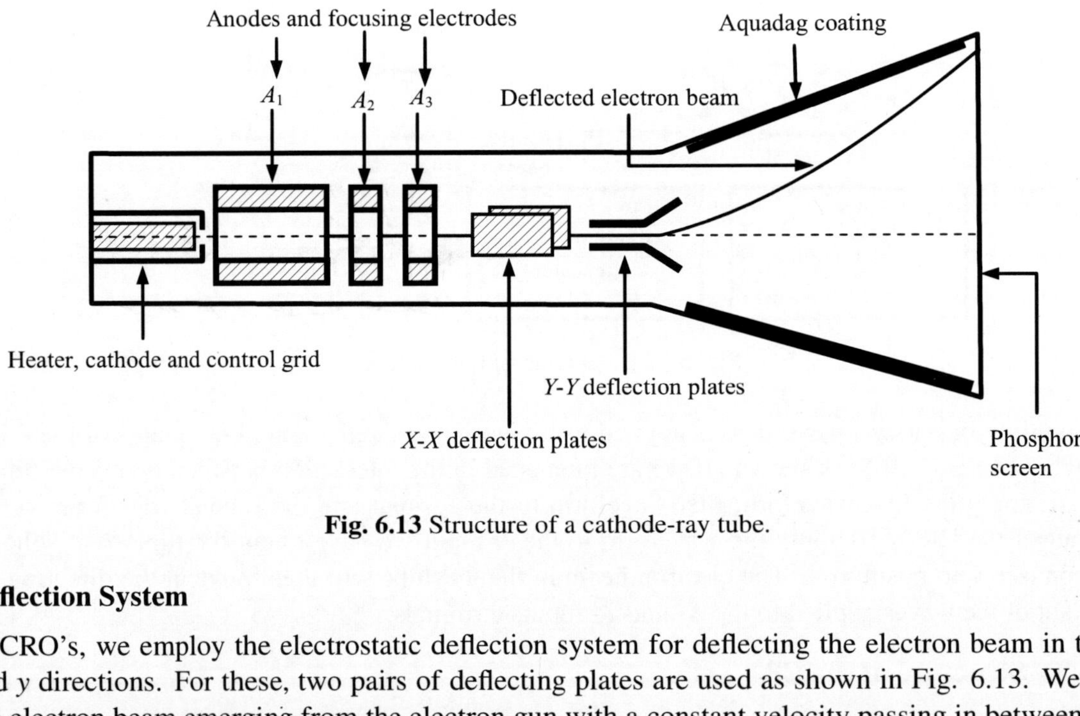

Fig. 6.13 Structure of a cathode-ray tube.

Deflection System

In CRO's, we employ the electrostatic deflection system for deflecting the electron beam in the *x* and *y* directions. For these, two pairs of deflecting plates are used as shown in Fig. 6.13. We find the electron beam emerging from the electron gun with a constant velocity passing in between two deflecting plates before striking the CRT screen.

Deflection Sensitivity

Deflection sensitivity of the CRO is defined as

$$S = D/V_{\mathrm{D}} = lL/2dV_{\mathrm{A}} \text{ meter/volt}$$

where D is the deflection produced on the screen, l is the length of the deflection plates, L is the distance between the plates and the screen, and d is the distance between the plates.

Deflection sensitivity represents the amount of deflection that the electron beam undergoes for one volt of applied deflection potential, and is a figure-of-merit of the CRT. The inverse of the sensitivity is called the *deflection factor G*. Thus, we find

$$G = 1/S = 2dV_{\mathrm{A}}/lL \text{ V/m}$$

Post-Deflection Acceleration (PDA) Scheme

We find that the deflection is inversely proportional to the final-anode voltage V_A and the distance between the plates d. Thus to get more deflection, we must apply small anode voltages.

Now, we come across a contrasting situation. We want the beam striking the screen to produce good brightness on the screen, for which its velocity must be large. This in turn requires that V_A must be high.

Also, with large deflecting voltages, rate of change of voltage (i.e., dV/dt) and hence the charging current CdV/dt, where C is the parasitic capacitance between the deflection plates, becomes very high. This is a dangerous condition.

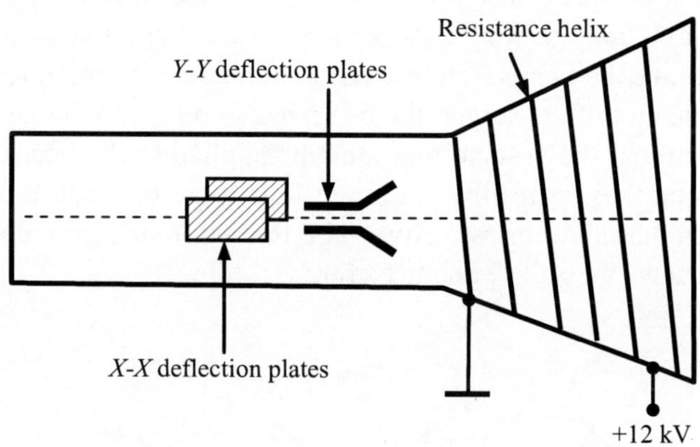

Fig. 6.14 Post-deflection acceleration system.

To avoid the problems mentioned above, in modern oscilloscopes, we first use small anode voltages so that the velocity of the electron entering the deflection region is small. After the beam has been deflected by these small voltages, we accelerate the beam so that it strikes the screen with a much higher velocity. *The acceleration so employed after defection is known as post-deflection acceleration.* Figure 6.14 shows the structure of a post-deflection acceleration system.

In this scheme, a resistive material is deposited inside of the CRT on its conical portion after the last pair of deflection plates in the form of a helix, as shown. This helix is kept at ground potential near the deflection plates and at +12 kV near the screen end of the tube. This large potential will provide constant acceleration to the electrons in the beam so that it will strike the screen with sufficient velocity to produce the desired brightness on the screen.

CRT Screen

Certain compounds like zinc sulphide exhibit the property of fluorescence, which a material emits light when an electron beam strikes it. The glass screen of a CRT is coated inside with a fluorescent coating. The special requirement for this coating is that it has to exhibit the property of phosphorescence also, a property of fluorescent materials by which light is emitted for some more time after it has been produced. This is called *persistence* of phosphor.

Phosphors are usually bad electrical conductors. This may lead to the deposition of residual charge on the screen, effectively decreasing the energy of the impacting electrons due to electrostatic repulsion (an effect known as "sticking"). To eliminate this, a thin layer of aluminium is deposited over the phosphors and connected to the conductive layer inside the tube. This layer also reflects the phosphor light to the desired direction, and protects the phosphor from ion bombardment resulting from an imperfect vacuum.

Time-Base Voltage Generator

The deflected electron beam striking the screen produces the desired waveform. For the generation of regular waveforms, as stated previously, we must apply a time-base voltage across the horizontal deflection plates. We know that the time-base voltage is a voltage that is increases linearly with time. This voltage will move the beam from the left extreme of the screen to its right extreme. On reaching the right extreme end of the screen the beam has to return to the left extreme to begin the next sweep. Thus we find that the voltage that must be applied to the horizontal deflection plates is a ramp voltage that increases with time, and then falls to zero. Such a waveform is shown in Fig. 6.15. It can be seen that a sweep waveform is a linearly rising and abruptly falling periodic waveform. Figure 6.15 shows the AC sweep waveform.

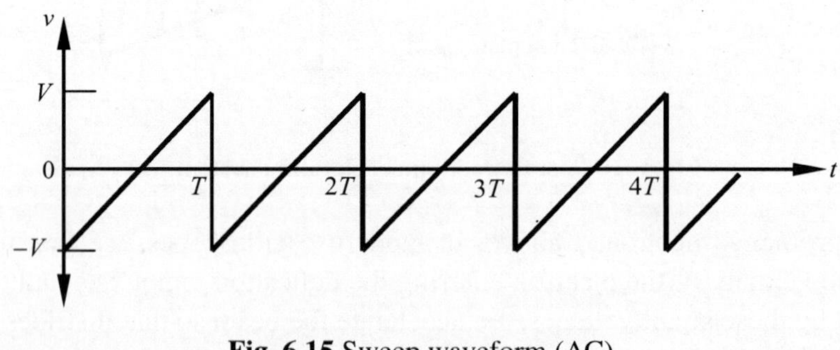

Fig. 6.15 Sweep waveform (AC).

Aquadag Coating

When electron beam strikes the phosphor screen, secondary electrons are emitted from the surface of the screen, and they get accumulated there unless they are removed. When their accumulation becomes quite large, they start repelling the electron beam away from the screen, which will create distorted images on the screen. To prevent the above problem, in all modern CRO's, a conductive graphite coating, called *aquadag*, is made on the inner wall of the flared end (or neck) of the CR tube, as shown in Fig. 6.13. This coating is also kept at high positive potential as the accelerating anode. This will perform two functions:

- Since the aquadag coating is positive, it attracts the secondary electrons and removes them.
- Since the aquadag is mounted in front of the accelerating anode, it aids in the acceleration of electrons toward the screen.

6.12.3 Applications of CRO

(a) **Observation of waveforms:** To observe these waveforms, we apply the waveform to be observed to the vertical input terminals. The time-base voltage is automatically applied by the internal time-base generator to the horizontal inputs, and this results in the waveform being displayed on the screen.

(b) **Measurement of voltage:** The amplitude of an applied voltage can be measured directly using the volt/cm-setting on the CRO's face.

(c) **Measurement of current:** CRO's are basically voltage-operated instruments. Therefore, to measure current, we must use a known resistor across which the current to be measured will develop a voltage. Then we measure this voltage using the CRO and compute the current as $I = V/R$.

(d) **Observation of Lissajous figures:** Lissajous figures were first studied by Nathaniel Bowditch, an American in 1815. They were studied independently in more detail by Jules-Antoine Lissajous in 1857. Lissajous figures are mathematical curves that results in when two independent forces are compounded. For example, consider two forces represented, respectively, by $x = V_x \cos \omega t$ and $y = V_y \cos \omega t$. Then dividing y by x, we obtain

$$y/x = V_y \cos \omega t / V_x \cos \omega t = V_y / V_x \quad (6.30)$$

If $V_x = V_y$, then Eq. (6.30) gets modified as

$$y = x \quad (6.31)$$

and this represents a straight line inclined at 45° with the x-axis. This straight line is the Lissajous figure that represents the resultant, when we compound two equal forces of the same amplitude and frequency, as shown in Fig. 6.16(a).

Now suppose, we make $x = V_x \cos \omega t$ and $y = V_y \sin \omega t$. Then we find

$$(x/V_x)^2 + \left(y/V_y\right)^2 = \cos^2\omega t + \sin^2\omega t = 1$$

The resultant figure thus is an ellipse, as shown in Fig. 6.16 (b). In the special case when $V_x = V_y$, the ellipse turns out to be a circle defined by [Fig. 6.16(c)]

$$x^2 + y^2 = 1$$

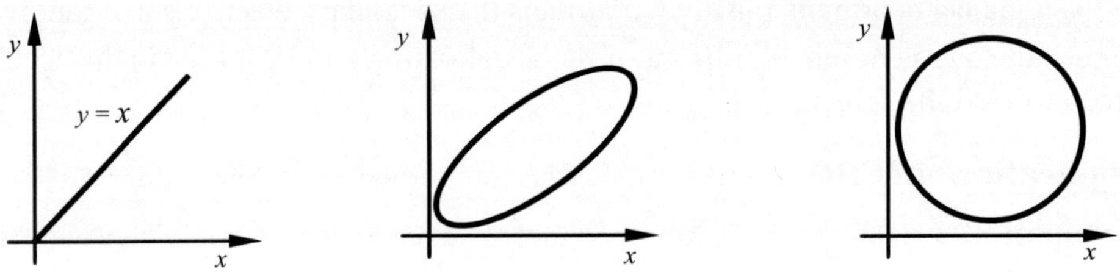

a. $y = x$ **b.** $(x/V_x)^2 + (y/V_y)^2 = 1$ **c.** $x^2 + y^2 = 1$

Fig. 6.16 Lissajous figures.

(e) **Measurement of unknown frequency:** Let the forces be represented, respectively, by $x = V_x \cos \omega_x t$ and $y = V_y \cos \omega_y t$, where the frequencies ω_x and ω_y are different, then the resultant Lissajous waveforms show complex shapes depending on their values. For example, if $\omega_y = 2\omega_x$, then the shape shown in Fig. 6.17(a) is developed as the resultant. It can be proved that the frequencies are related to the each other as

$$\frac{\omega_x}{\omega_y} = \frac{\text{No. of tangent points along the vertical direction}}{\text{No. of tangent points along the horizontal direction}}$$

where the tangent points are those that are obtained when tangent lines are drawn in the vertical and horizontal directions as shown in Fig. 6.17(a). Figure 6.17(b) shows a similar situation when

$$\omega_x/\omega_y = 4/1$$

It can be seen that by using a standard frequency as ω_x, we can determine ω_y using the figure obtained when we compound the frequencies on a CRO screen. To use the CRO for this purpose, we apply the voltage $x = V_x \cos \omega_x t$ to the horizontal inputs and the voltage $y = V_y \cos \omega_y t$ to the vertical inputs. It is important to note that when we apply an external voltage to the X-X input terminals, the internal sweep (time-base) voltage is to be switched off.

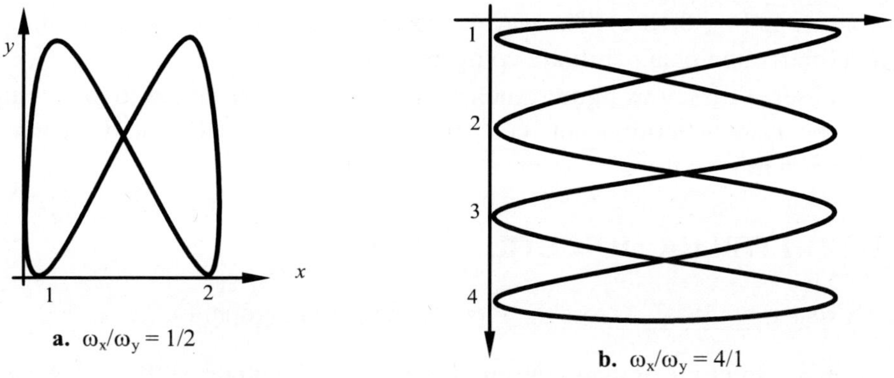

Fig. 6.17 Computation of unknown frequency.

6.12.4 Typical CRT Connections

Figures 6.18 show some typical connection methods used in CRO's. Figure 6.18(a) shows the essential controls on the face panel of the scope. Figure 6.18(b) shows how to apply an input to the *Y* terminal of a CRO. The waveform to be observed is always applied to the *Y*-input terminal of the CRO, if it is a single-beam scope. If it is of the dual-trace or of the dual-beam type, then the input will be applied through one of the *Y*-input terminals. By suitably adjusting the **Time/div** control and the **Volt/div** controls, we can observe the waveform on the CRT screen. By using these controls, we can measure the frequency and amplitude, respectively, of the applied wave.

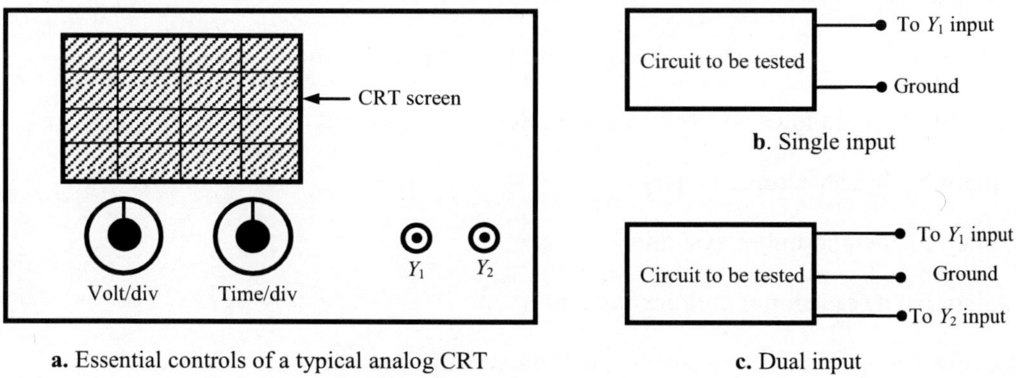

Fig. 6.18 Typical CRO connections.

If we are trying to find the frequency response of an amplifier, then we have to observe the input and output waves at the same time. This is possible by using the dual-trace or beam type CRT. To illustrate the idea, consider an experiment in which we have to obtain the frequency response of an amplifier. In this case, the input and output waves of the amplifier are applied to the Y_1 and Y_2 inputs of the CRO, as shown in Fig. 6.18(c). The waveforms are arranged to be one below the other so that their relative amplitudes and phase shift are compared.

If we have to observe Lissajous figures, then we have to apply one input to the *X*-input terminal and the other to the *Y*-input terminal of the CRO. By controlling the Volt/div controls, we can observe the required figure on the CRT screen.

ONE-WORD/MULTIPLE-CHOICE QUESTIONS

M6.1 There exists a ________ across the input terminals of an opamp.

M6.2 An inverting amplifier is one that produces ________ phase shift between its input and output voltages.

M6.3 A non-inverting amplifier is one that produces ________ phase shift between its input and output voltages.

M6.4 The gain of an inverting amplifier is ________.

M6.5 The gain of a non-inverting amplifier is ________.

M6.6 An opamp shorted between its inverting terminal and output terminal is called ________.

M6.7 An opamp integrator has a ________ resistor connected across its input/output terminals.

REVIEW QUESTIONS

1. Explain different types of number systems. Illustrate with suitable examples.
2. With suitable examples, show how one system can be converted into another.
3. Explain the binary number system.
4. Explain the octal number system.
5. Explain the hexadecimal number system.
6. Convert decimal number to binary and binary to decimal number.
7. Convert decimal number to octal and octal to decimal number.
8. Convert decimal number to binary hexadecimal and hex to decimal number.
9. Explain binary fraction.
10. Convert decimal fraction to binary fraction.

11. Discuss binary codes used in communication systems.
12. Explain BCD codes. Specify its advantages and disadvantages.
13. Compare binary and BCD numbers.

UNSOLVED PROBLEMS

P6.1 If the input voltages to an opamp summer are 1, 1, 1, and 1, respectively, and the respective resistors are 1 kΩ, 2 kΩ, 4 kΩ, and 8 kΩ, and the feedback resistor is 8 kΩ, compute the value of the output voltage of the summer.

P6.2 If the input voltages to an opamp inverter is 10 V, determine the value of its output voltage.

P6.3 If the input voltages to an opamp integrator is 10 sin ($200\pi t$) V, determine value of its output voltage. Assume suitable values of R and C.

P6.4 Find the output voltage of an opamp differential amplifier if the input voltages are 2 V and 4 V, respectively and the resistors used are $R_1 = R_3 = 10$ kΩ and $R_2 = R_4 = 100$ kΩ.

CHAPTER 7

COMMUNICATION SYSTEMS

7.1 RADIO COMMUNICATION SYSTEMS

In 1864, James Clark Maxwell predicted the existence of electromagnetic waves, and in 1887, Heinrich Hertz invented a method to produce these waves. In 1898, Macroni invented a method to transmit and receive electromagnetic waves, which later came to be known as *radio waves*.

To transmit sound waves through space, we first convert them to corresponding electric signals by using microphones. These electric signals are then fed to an antenna, which radiates them into space as electromagnetic waves.

7.1.1 Modulation

We define modulation as *the process of varying one of the parameters (amplitude, frequency, phase, intensity, or polarisation) of a higher frequency wave called the carrier wave, in accordance with the instantaneous amplitude of a lower frequency called the modulating wave.*

From the above general definition, we may define *amplitude modulation as the process of varying the amplitude of the carrier wave in accordance with the instantaneous amplitude of the modulating wave.* In modulation, it is imperative that the carrier frequency f_c is to be much higher than the modulating frequency f_m.

7.1.2 Need for Modulation

Let us now investigate on why we need modulation. The following points highlight the necessity of using various modulation techniques.

- To transmit electromagnetic waves efficiently, it is found that the antenna to be used must have a height h, given by the equation:

$$h = \frac{\lambda}{2} \tag{7.1}$$

  where λ = wave length of the radiation used. Electromagnetic frequencies in the audio range exist between 30 Hz and 30 kHz. As such, the corresponding wavelengths range from 10^4 meters to 10^7 meters. It is practically extremely difficult to construct antennas having these large heights.

- Different transmitting stations must have different frequencies for transmission; otherwise it will be difficult to receive programs from various stations separately.
- The higher the modulating frequency, the more the power in the transmitted waves, and the less the transmitter power required. For example, a 1000-kW transmitter is a high-power transmitter in the medium-wave range, whereas a 10-kW transmitter is a high-power transmitter in the very high-frequency range. At microwaves, a 100-W transmitter is considered as a high-power transmitter.
- Using single-side band systems, several audio/video channels can be transmitted over a single cable using microwave, or optical frequencies.
- Using modern digital modulation (pulse-code modulation) techniques, noise can be reduced greatly.
- Using various coding schemes, secrecy in transmission is possible.
- Using separate coding schemes, even signals that are considered to be lost in background noise can be recovered.
- Using data-compression techniques, economy in transmission is possible.

There are several other advantages of using digital modulation schemes; but describing them is beyond the scope of this text book.

7.1.3 Amplitude Modulation

In amplitude modulation (AM), as stated before, the amplitude of the carrier wave is varied as per the instantaneous amplitude of the modulating wave. To develop an analytical expression for AM, let the carrier wave be represented by:

$$v_c = V_c \cos \omega_c t \tag{7.2}$$

where v_c = instantaneous amplitude, V_c = peak value, and ω_c = frequency of the carrier wave. In a similar way, let the modulating wave be represented by:

$$v_m = V_m \cos \omega_m t \tag{7.3}$$

where v_m = instantaneous amplitude, V_m = peak value, and ω_m = frequency of the modulating wave. After modulation, the carrier amplitude V_c is varied in accordance with the instantaneous amplitude v_m of the modulating wave. In mathematical form, the amplitude of the modulated wave is given by:

$$V_c + V_m \cos \omega_m t \tag{7.4}$$

In Eq. (7.4), we have added the modulating wave as such to the carrier amplitude. This means that the carrier amplitude will vary according to the instantaneous value of the modulating wave. The expression of the modulated wave is now given by:

$$v_t = (V_c + V_m \cos \omega_m t) \cos \omega_c t \tag{7.5}$$

Equation (7.5) may be expanded as:

$$v_t = V_c\left[1 + \frac{V_m}{V_c}\cos\omega_m t\right]\cos\omega_c t = V_c\,[1 + m\cos\omega_m t]\cos\omega_c t$$

$$= V_c\cos\omega_c t + mV_c\cos\omega_c t\cos\omega_m t$$

$$= V_c\cos\omega_c t + \frac{mV_c}{2}\cos(\omega_c + \omega_m)\,t + \frac{mV_c}{2}\cos(\omega_c - \omega_m)\,t \quad (7.6)$$

In Eq. (7.6), $m = V_m/V_c$ is called the *modulation index.* It represents the ratio of the peak amplitudes of the modulating and the carrier waves.

Equation (7.6) shows that the modulated wave contains three frequency components, viz., f_c, $(f_c + f_m)$ and $(f_c - f_m)$. Let us, for example, assume $f_c = 1000$ kHz, and $f_m = 1000$ Hz. Then, after modulation, the frequency components in the modulated wave are 1000 kHz (f_c), 1001 kHz ($f_c + f_m$) and 999 kHz ($f_c - f_m$), respectively. So, we find that the frequency of the modulating wave has indeed been raised. However, we also find that the modulated wave contains the modulating frequencies along with the carrier frequency. These modulating frequencies can be recovered from the modulated waves by means of demodulation techniques.

7.1.4 Waveforms associated with Amplitude Modulation

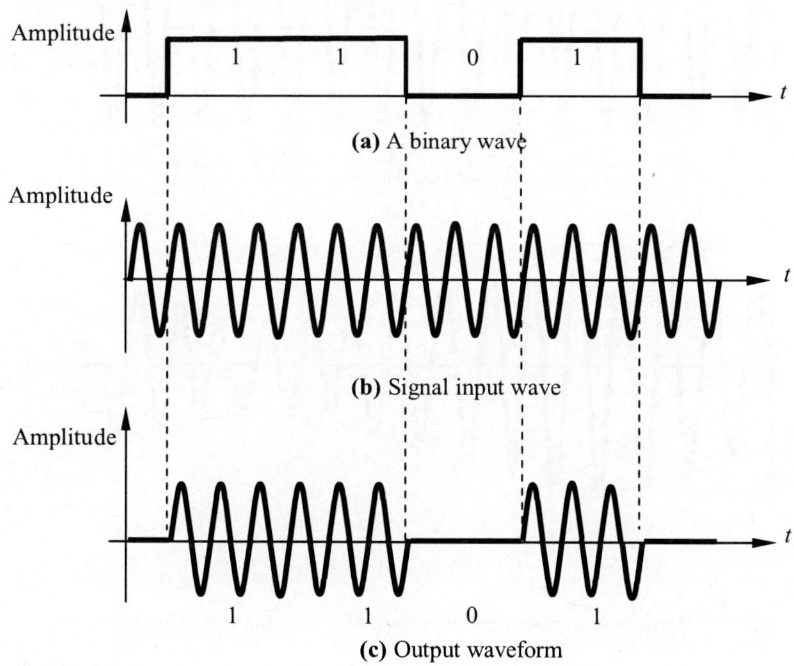

Fig. 7.1 Waveforms associated with amplitude modulation using square wave.

Consider a torchlight producing an optical output. This may be considered as a sinusoidal oscillation. When the light is turned off, the output will be zero. Now, consider Fig. 7.1(a), which shows the ON-OFF light conditions, represented by sinusoidal signals. We find that this complex waveform, shown in Fig. 7.1(c), may be assumed to be composed of the two waveforms shown in Figs. 7.1(a) and (b), respectively.

Figures 7.2 show waveforms associated with amplitude modulation using sinusoidal modulating waves. Figure 7.2(a) shows the sinusoidal modulating wave and Fig. 7.2(b) shows the sinusoidal carrier wave. It may be noted that the frequency of the carrier wave is much higher than that of the modulating wave. Figure 7.2(c) shows the resultant amplitude-modulated waveform. It can be seen from Fig. 7.2(c) that the carrier amplitude varies at every instant in accordance with the instantaneous amplitude of the modulating signal.

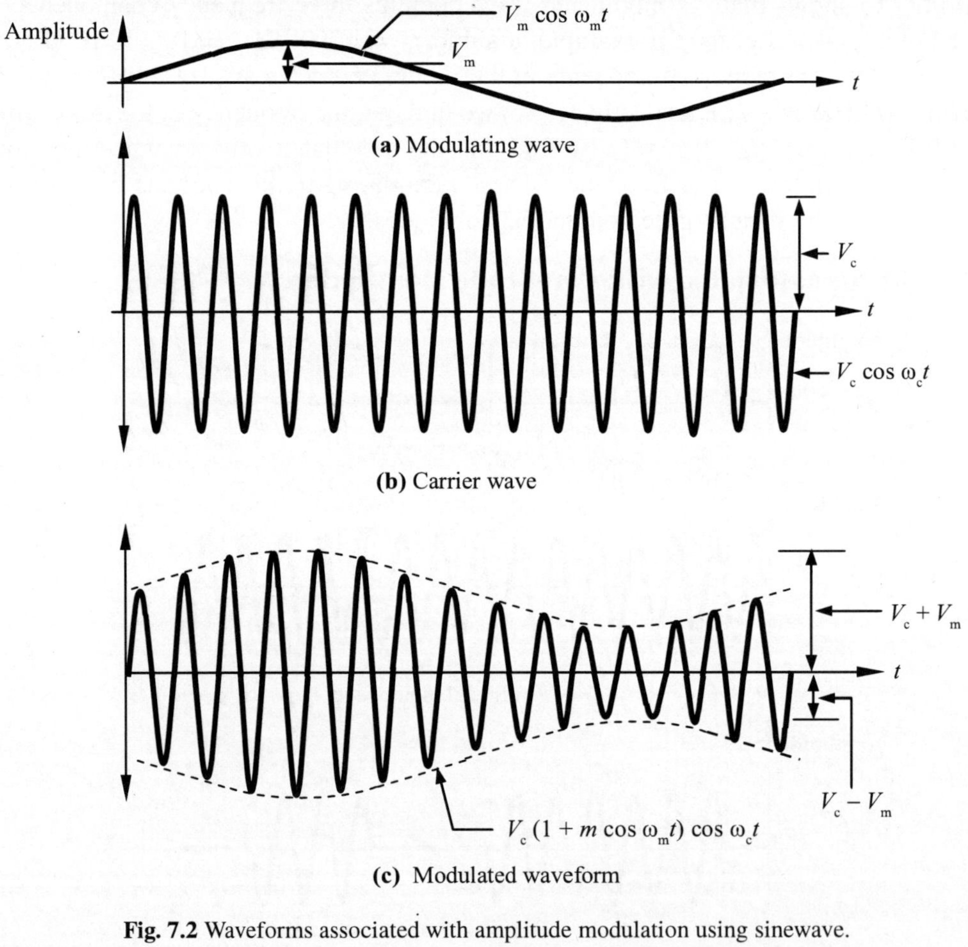

Fig. 7.2 Waveforms associated with amplitude modulation using sinewave.

7.1.5 AM Transmitter

AM transmitters are of two types:

- Low-power transmitters (LPT's)
- High-power transmitters (HPT's).

LPT's are modulators with power levels less than a few kilowatts. HPT's are transmitters whose power levels exceed 10 kW. At present, LPT's are constructed by using semiconductor devices. However, HPT's are constructed by using vacuum tubes. This is one of the areas, where vacuum tubes are still in use. Vacuum tubes are used for high-power modulators because of the non-availability of power transistors that can handle high voltages and currents.

Low-Power Transmitters (Modulators)

Figure 7.3 shows the block schematics of a low-power (level) transmitter (modulator). As shown in the figure, an LPT consists of a master oscillator, an isolation amplifier, a modulator, and an RF power amplifier (or a group of power amplifiers) in the radio-frequency section. The audio-frequency section consists of the microphone, an AF amplifier, and an AF power amplifier. The AF power amplifier output is applied to the modulator, as shown. It may be noted that the microphone can be replaced with a video camera. Then by appropriately modifying the amplifiers, we get a television transmitter in this case. The blocks are now explained in detail, as given below.

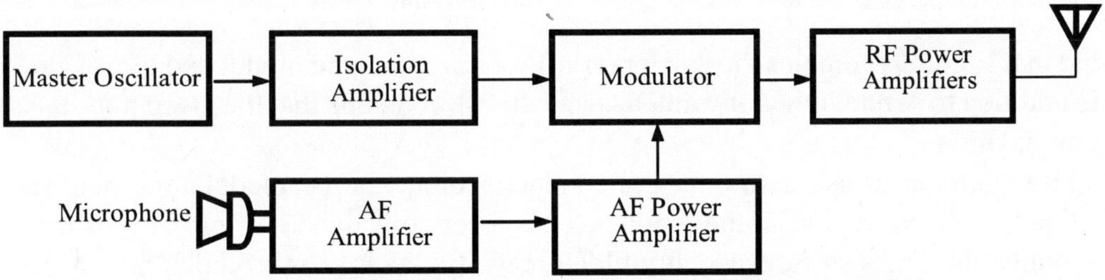

Fig. 7.3 Block diagram of low-level modulator.

Master Oscillator

The most important part of any transmitter is the master oscillator. It is the master oscillator that produces the carrier (transmission) frequency f_c of the transmitter. This is the frequency to which a receiver must be tuned to so that programs from a given station can be received in. This necessitates the fact that the master-oscillator output frequency must be highly constant and extremely stable. To get a highly constant frequency, we use crystal oscillator as the master oscillator. We know that a crystal produces a highly stabilised frequency when operated in a temperature-controlled atmosphere. It may also be noted in this case that the master-oscillator frequency will usually be a sub-multiple of the actual frequency to be transmitted. For example, if the actual carrier frequency to be transmitted is 1 MHz, then we may use a master oscillator that produces a frequency of only 100 kHz.

Isolation Amplifier

Master oscillators are to be isolated from the loading effects of succeeding stages. For this, isolation amplifiers are used. Emitter-followers and class-*C* amplifiers can be used as isolation amplifiers. Emitter-follower is an amplifier with its input at the base and output at the emitter. This amplifier configuration has high input impedance so that it will not load its previous stage, viz., the master oscillator, and a very low output impedance so that its succeeding stage (modulator) will not load it. Thus, we find that emitter-follower is a good isolation amplifier.

Modulator

A modulator is a non-linear circuit, which produces an output according to the relation:

$$v_o = av_i + bv_i^2 + cv_i^3 + ... \tag{7.7}$$

where v_i = input voltage, v_o = output voltage, and $a, b, c, \ldots$ are constants. Now, consider the case when the input

$$v_i = V_c\cos \omega_c t + V_m\cos \omega_m t \tag{7.8}$$

Assuming $v_o = v_i + v_i^2$, and neglecting all other terms we find, using Eqs. (7.7) and (7.8), that

$$\begin{aligned} v_o &= (V_c\cos \omega_c t + V_m\cos\omega_m t) + (V_c\cos \omega_c t + V_m\cos \omega_m t)^2 \\ &= V_c\cos\omega_c t + V_c\cos \omega_m t + V_c^2\cos^2\omega_c t + V_m^2\cos^2\omega_m t + 2V_c V_m\cos \omega_c t \cos \omega_m t \end{aligned} \tag{7.9}$$

We find that Eq. (7.9) contains the carrier term $V_c\cos \omega_c t$, and the modulated term $V_c V_m\cos \omega_m t \cos \omega_c t$, in addition to several other unwanted terms. So we conclude that the given non-linear circuit produces modulation.

It is quite common to use a simple semiconductor diode as the modulator. A diode can be operated in its non-linear region, if the input exceeds a certain value. This property is made use of in diode modulators. Transistors, especially FET's, can also be used as modulators.

AF (Video) Amplifier and Power Amplifier

To the modulator, the second input other than the carrier-frequency input, is the modulating audio (or video) input frequency. This is obtained as a separate input by using a microphone (or video camera). The output of the microphone (or video camera) is sufficiently amplified before applying it to the modulator.

Final RF Power Amplifier

This is usually a class-*B* or class-*C* power amplifier. The purpose of this amplifier is to bring the power of the modulated signal to the required level before being applied to the transmitting antenna.

High-Power Modulators

Figure 7.4 shows the block diagram of a high-level modulator. The structure of LPT's and HPT's are more or less similar, except that in, the power-levels involved are very high. As stated before, radio transmitters having power outputs greater than 10 kW can be considered as HPT's. From Fig. 7.4, we find that modulation of AF and RF signals is done at high power levels.

We also notice that in HPT's also, the master oscillator is designed to produce a sub-multiple frequency of the final carrier frequency. This is then frequency-multiplied by a series of class-C power amplifiers, which also act as frequency multipliers. Class-C amplifiers are tuned amplifiers, which can be used as RF voltage and power amplifiers by appropriately biasing the transistors/vacuum tubes used. When used as frequency-multipliers, class-C amplifiers are called *harmonic generators.* It may be noted that harmonic generators multiply the fundamental frequency, as well as raise the power level of the signal. To modulate the high-power RF carrier, we must also raise the power level of the modulating frequency using several stages of amplification, as shown in Fig. 7.4.

As in the case of LPT, in HPT also, the final stage after modulation is a class-B or class-C power amplifier, which will raise the power of the modulated signal to the level required for transmission through the antenna.

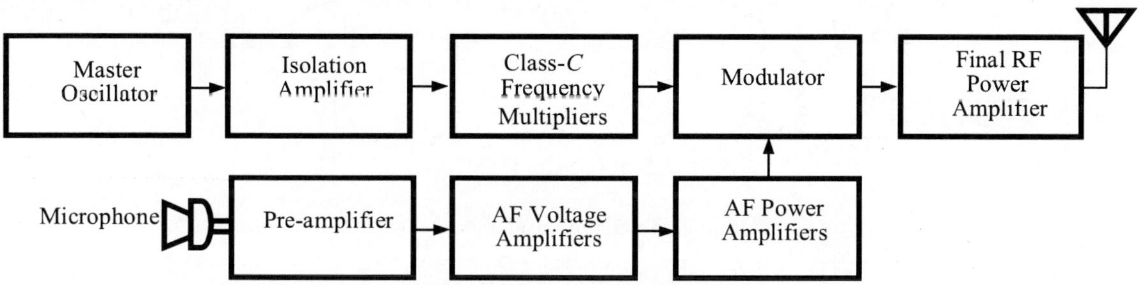

Fig. 7.4 High-level modulators.

7.1.6 AM Radio Receivers

The process of recovering the original (modulating) signal from the modulated signal is known as demodulation or detection. The basic principle can be explained with the help of a crystal detector.

Simple crystal detector

Figure 7.5 (a) shows a simple crystal receiver. The crystal receiver consists of a crystal detector diode D, which is driven by a tuning circuit consisting of an RF transformer and capacitors C_1 and C_2. crystal diode is the old name for ordinary point-contact diodes, such as OA79 and 1N34. Capacitors C_1 and C_2 are variable capacitors of the air- or plastic-core type, ganged together, so that they can be rotated on a common spindle. Because of the ganging, they will have the same capacitance value at any setting of the spindle. These capacitors are available in the open market under the trade name PVC 2J, whose value can be varied between 30 pF and 330 pF.

The simple crystal-diode receiver also contains a filter capacitor C, and a load resistance in the form of a pair of headphones. The working of the receiver may be explained as follows:

The RF signal from a given radio station is tuned in by gang capacitors C_1 and C_2 and the primary and secondary windings of the RF transformer, according to the expressions:

$$f_o = 1/2\pi\sqrt{L_1C_1} = 1/2\pi\sqrt{L_2C_2} \tag{7.10}$$

where f_o = frequency of the radio station to be tuned in, L_1 = inductance of the primary, and L_2 = inductance of the secondary of the RF transformer. As the tuned-in modulated radio-frequency signal is AC, the diode, which acts as a rectifier, converts this AC into its corresponding DC. The rectified DC waveform is shown in Fig. 7.5 (b).

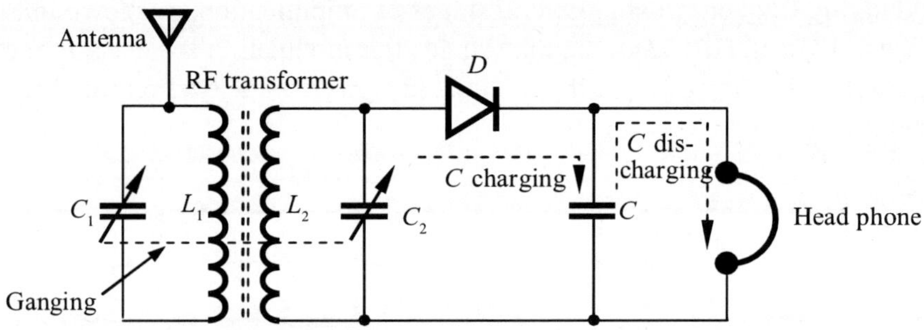

Fig. 7.5(a) Crystal receiver.

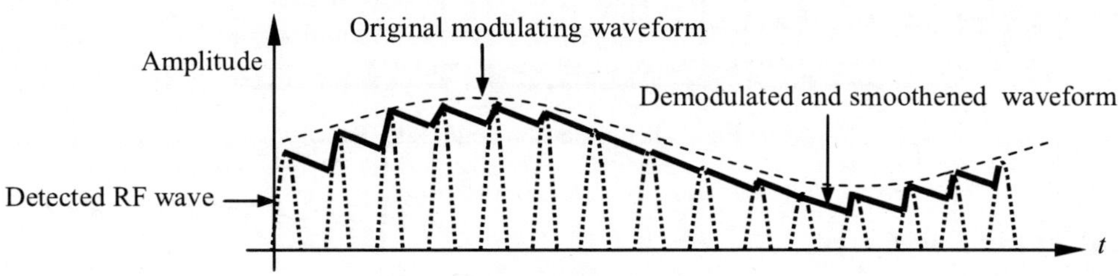

Fig. 7.5(b) Detected wave.

As shown in Fig. 7.5(a), the rectified DC is applied then to a capacitor C. The capacitor gets charged to the peak of each of the rectified DC half-cycle (i.e., positive half-cycle of the AC). During the negative half-cycles, when the diode is off, C gets discharged through the headphone, which acts as the load resistance of the detector. The charging and discharging of C results in the modulating waveform being reproduced through the headphone.

Figure 7.5(b), as stated before, shows the rectified and filtered waveform. The dotted lines in the figure represent the rectified DC output of the diode. The solid lines joining the tips of the dotted-line half-waves represent the waveform associated with the charging and discharging of C. The smooth waveform above the solid-line waveform, drawn using dotted lines again, represents the actual modulating signal. We find that the waveform generated due to the charging and discharging of C approximately resembles actual modulating signal. The differences between these two signal waveforms represent noise and distortion.

Advantages of the simple crystal receiver

- The theory of the crystal receiver is very simple and straightforward.
- It can be constructed with a few components.

Disadvantages of the simple crystal receiver

- The tuning circuit is not perfect; this results in the most powerful local station being tuned in.
- Only the most powerful local station can be received. Other low-power local stations can be heard in the background.
- Received power level is very low. So, the signal can be heard only through a headphone. This means that only one person can hear the station tuned in.
- Filtering is imperfect. Hence noise and distortion are high.

Superheterodyne Radio Receiver

The defects of the simple crystal receiver is rectified in the superheterodyne radio receiver, invented by Major Edwin Armstrong of thc US army. The block diagram of the superheterodyne radio receiver is shown in Fig. 7.6.

The signal is tuned in by the antenna and associated tuning circuits, as in the case of the simple crystal receiver. The signal so tuned in is applied to an RF amplifier, which acts to amplify the low-level RF input fed-in by the antenna. RF amplifier is used for the following purposes:

- To act as a booster amplifier to improve the signal-to-noise ratio.
- To provide a good coupling between the antenna and the input stage of the receiver.
- To provide proper impedance matching between the antenna and the receiver.
- To provide good adjacent channel selectivity.
- To provide image-frequency rejection.
- To prevent re-radiation of signals from the local oscillator to avoid echo in radios.
- To improve the sensitivity of the receiver by amplifying very low-level signals.

The RF amplifier is a class-*A* tuned voltage amplifier. The tuning is done by using a gang capacitor, similar to the one we have used for the crystal detector.

The output of the RF amplifier is applied to a non-linear circuit called the mixer, to which we apply the output of the local oscillator also. The local oscillator is an LC oscillator of the Hartley, Colpitt's, or Armstrong type. This oscillator is designed to generate a frequency that is greater than the incoming signal frequency by a value known as the intermediate frequency (IF). For medium-wave operation (i.e., frequency range between 500 kHz and 700 kHz), the IF is chosen to be 455 kHz.

The mixer is a non-linear circuit, which will produce an output that contain the sum and difference of frequencies applied to it. This means that we will get the sum and the difference of the local

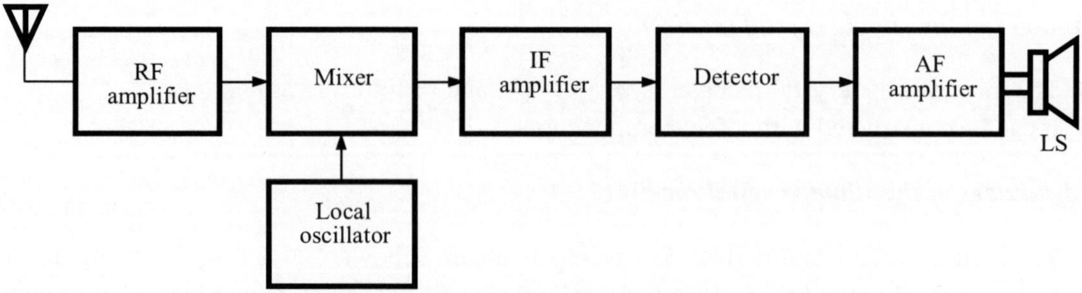

Fig. 7.6 Block diagram of the superheterodyne receiver.

oscillator frequency (f_o) and the incoming signal frequency (f_s). We design the IF amplifier such that it is tuned to select only the difference of f_o and f_s. That is:

$$f_o - f_s = f_i \tag{7.11}$$

We call the difference frequency f_i as the intermediate frequency (IF). The advantage of this is that each stage of the IF amplifier, since it is a single-frequency amplifier, can be designed to produce the maximum gain possible by that stage. Thus, in any radio receiver, it is the IF amplifier stage that produces the maximum gain in that receiver. We find that the IF amplifier converts any incoming frequency to f_i. It may be noted that, *in this conversion process, it is only the modulated frequency that gets converted to the IF; the amplitude of the modulated signal does not get changed.* This means that the shape of the envelope of the modulated voltage will not get changed at all in the IF conversion, which further means that the shape and frequency of the modulating voltage will remain as such.

The IF amplifier is, again, similar to the RF amplifier circuit. However, it is quite common to have two stages of IF amplifier in radio receivers using transistors.

The output of the IF amplifier is then applied to a diode detector, similar to the one we have seen in the case of the simple receiver. However, for the superheterodyne receiver, we use a pi-filter instead of the simple capacitor filter, for better filtering action. The rectified and filtered envelope of the modulated signal is nothing but the modulating voltage, and has a shape similar to that of the waveform shown in Fig. 7.5(b).

The output of the detector is in the AF range and of very low power level. This is therefore amplified by using an AF power amplifier, which drives a loud speaker. It may be noted that a power amplifier is usually a combination of a voltage amplifier and a current amplifier. The voltage amplifier enhances the low voltage level in the input demodulated signal to a high level, and the current amplifier enhances the low current level to a high level so that it can drive the loud speaker.

In addition to the above, a superhet receiver also contain automatic-volume-control (AVC) circuit, power-supply circuit, etc. But, these are not described here, as this is only an elementary treatment.

7.1.7 Frequency Modulation

In frequency modulation (FM), the frequency of the carrier gets modulated by the modulating voltage. The analytical expression for the FM can be derived as follows. Let

$$v_c = V_c \cos \omega_c t \tag{7.12}$$

$$v_m = V_m \cos \omega_m t \tag{7.13}$$

Then, after frequency modulation, the frequency of the modulated wave becomes

$$\omega_t = \omega_c + k_f V_m \cos \omega_m t \tag{7.14}$$

where k_f = a constant of proportionality. The phase of the modulated wave can be obtained by integrating Eq. (7.14). Performing the required integration, we get

$$\phi_t = \int_0^t \omega_t dt = \omega_c t + \frac{k_f V_m}{\omega_m} \sin \omega_m t = \omega_c t + m_f \sin \omega_m t \tag{7.15}$$

where $m_f = k_f V_m / \omega_m$ is the modulation index in FM. The expression for the modulated wave is now given by:

$$v_t = V_c \sin (\omega_c t + m_f \sin \omega_m t) \tag{7.16}$$

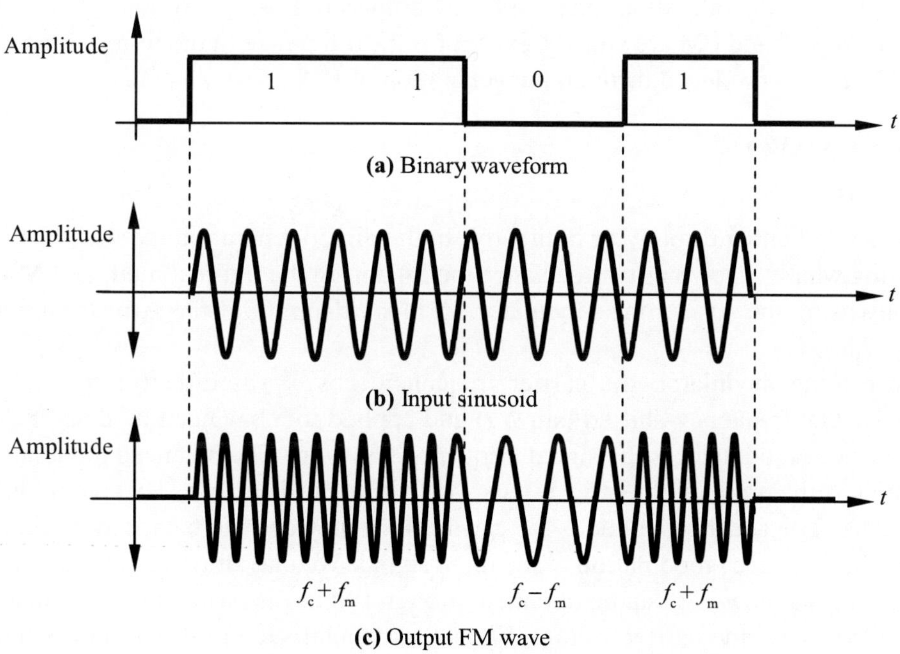

Fig. 7.7 Waveforms associated with frequency modulation.

Equation (7.16) represents a complex mathematical function, which can be expanded into a Bessel's series function. It can be seen from the Bessel's function expansion of Eq. (7.16) that the frequency-modulated wave contains infinite frequency components of the type $(f_c \pm f_m)$, $(f_c \pm 2f_m)$, $(f_c \pm 3f_m)$, etc. The waveforms associated with FM using a binary (digital) modulating wave are shown in Figs. 7.7.

As shown in Fig. 7.7 (c), the modulated wave contains the frequency components $(f_c + f_m)$ and $(f_c - f_m)$, respectively. Frequency components $(f_c + f_m)$ correspond to modulation by 1's, and frequency components $(f_c - f_m)$ correspond to modulation by 0's. If we use sinusoidal modulation, the modulated wave will contain a gradual increase in frequency corresponding to $+v_m$ and a gradual decrease in frequency corresponding to $-v_m$. But this waveform is not shown here.

7.1.8 Phase Modulation

In phase modulation (PM), the phase of the carrier gets modulated by the modulating voltage. The analytical expression for the phase modulation can be obtained as in the case of FM by noting that after modulation, the modulated phase is given by:

$$\phi_t = \phi_c + k_p V_m \sin \omega_m t \tag{7.17}$$

where ϕ_c = carrier phase, and k_p = constant of proportionality. The modulated wave is then given by

$$v_t = V_c \sin \left(\omega_c t + m_p \sin \omega_m t \right) \tag{7.18}$$

where $m_p = k_p V_m$ is the modulation index in PM. Comparing Eqs. (7.16) and (7.18), we find that the expressions for FM and PM are similar, except for the difference in the terms m_f and m_p. It then follows that FM can be produced through the generation of PM.

7.1.9 Generation of FM

Armstrong modulator

FM can be generated either directly or indirectly. In the direct-generation method, FM is produced directly by using what is known as the reactance modulator. In the indirect method, FM is produced through PM by using the Armstrong modulator. A block diagram of the Armstrong modulator is shown in Fig. 7.8.

In the Armstrong modulator, the carrier frequency $(\cos \omega_c t)$ is directly applied to an adder circuit. It is also 90° frequency-shifted $(\sin \omega_c t)$ and applied to a balanced modulator. The second input to the balanced modulator is the signal frequency $(\sin \omega_m t)$. The balanced modulator produces an output, which is the product of its two inputs, viz., $\sin \omega_c t$ and $\sin \omega_m t$. This output is the second input to the adder. The adder circuit adds the carrier frequency and the balanced-modulator output. This results in the frequency modulation of the input signal. We may draw vector diagrams as shown in Figs.7.9 to illustrate how FM can be obtained through PM in the Armstrong modulator.

Figure 7.9 (a) shows the carrier voltage $\boldsymbol{V_c}$ as a horizontal vector. It also shows the side-band vectors at a finite angle with $\boldsymbol{V_c}$, and their resultant vector $\boldsymbol{V_m}$ 90° out-of-phase with respect to $\boldsymbol{V_c}$.

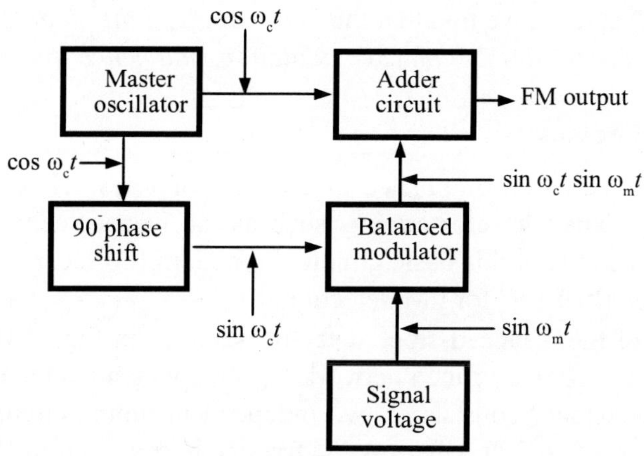

Fig. 7.8 Armstrong modulator.

Referring back to Fig. 7.8, we find that the carrier is phase-shifted through 90° before being applied to the balanced modulator (BM). The resultant of the side bands V_m generated by the BM can therefore be seen to be 90° out-of-phase with respect to the carrier. In Fig. 7.9 (a), we have shown that the amplitude of the modulating voltage small compared with that of the carrier. This produces a small resultant vector V_{m1}, as shown. The amplitude of the modulated voltage can be seen to be the vector sum of V_c and V_{m1}. This is designated as V_{R1} in Fig. 7.9 (a). We also notice that the phase angle between V_c and V_{m1} is designated as θ_1 and is comparatively small.

In Fig. 7.9(b), we find that V_{m2} is larger than V_{m1} as the amplitude of the modulating voltage is comparatively larger than that in the previous case. Hence we find that the modulated amplitude $V_{R2} > V_{R1}$, with a corresponding phase angle θ_2, which is larger than θ_1.

In Fig. 7.9(c), we have shown a modulating vector V_{m3}, which is greater than V_{m2}, resulting in $V_{R3} > V_{R2}$, and a corresponding phase angle $\theta_3 > \theta_2$.

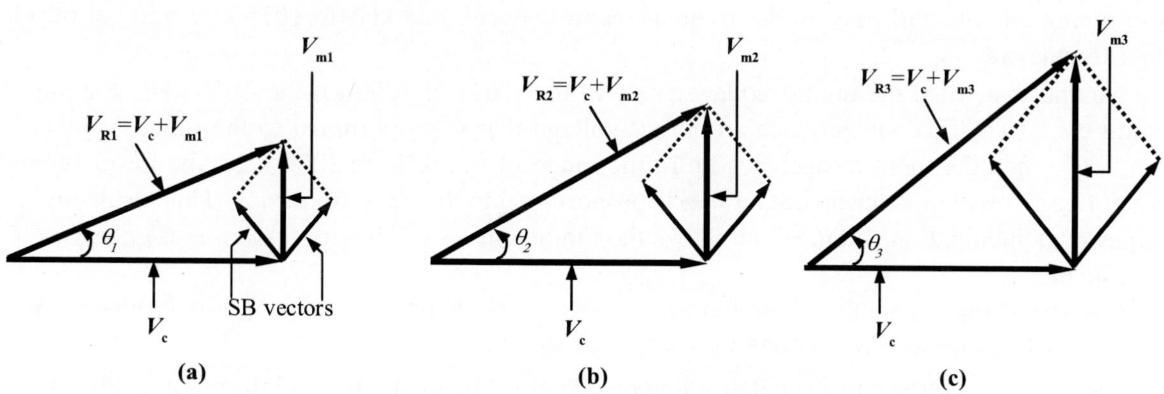

Fig. 7.9 Vector diagrams of FM-through-PM.

From Figs. 7.9(a), (b) and (c), we find that the *phase angle of the modulated vector varies as its amplitude. This operation is nothing but phase modulation, and hence frequency modulation.*

7.1.10 Detection of FM waves

Frequency demodulation can be accomplished in a variety of ways. Here, we describe the basic balanced-slope detector. Other detector circuits such as the Foster-Seely descriminator, and the Ratio detector are modifications of this basic circuit. In modern FM receivers, however, we mostly employ phase-locked loop (PLL) IC for FM detection.

The circuit diagram of the balanced-slope detector is shown in Fig. 7.10. The balanced-slope detector circuit consists of a primary-tuned network (L_1-C_1), which is tuned to frequency f_c of the modulated wave. The secondary consists of two independent tuning circuits (L_2-C_2 and L_3-C_3). Network L_2-C_2 is tuned to ($f_c + \Delta f$), where Δf = frequency deviation in FM. Similarly, network L_3-C_3 is tuned to ($f_c - \Delta f$). The operation of the circuit can be explained by considering practical values.

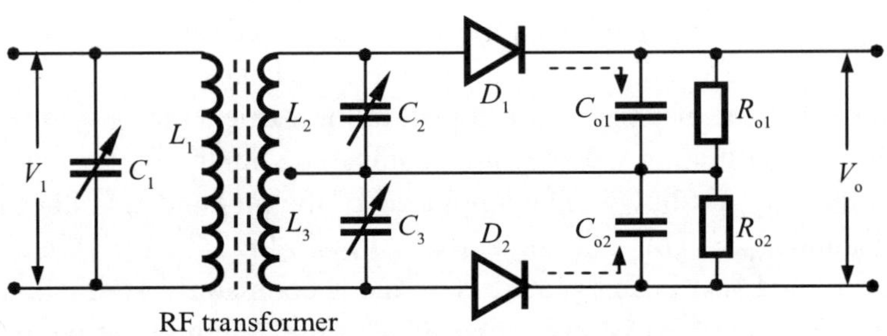

Fig. 7.10 Balanced-slope detector.

Let f_c = 1000 kHz, and Δf = 75 kHz. Then L_1-C_1 will pass all the frequencies in between 925 kHz to 1075 kHz. Now, we tune L_2-C_2 to 1075 kHz, and L_3-C_3 to 925 kHz. This means that both tuning circuits will pass all the frequencies in between 925 kHz to 1075 kHz and cut off all other frequencies.

We find that when the input frequencies lie in the range of 1000 kHz to 1075 kHz, the upper tuning circuit (L_2-C_2) will produce an output voltage that is proportional to the input frequency. Similarly, when the input frequencies lie in the range of 925 kHz to 1000 kHz, the lower tuning circuit (L_3-C_3) will produce an output that is proportional to the input frequency. This results in the frequency demodulation. Figure 7.11 shows the tuning curves of circuits L_1-C_1, L_2-C_2 and L_3-C_3, respectively.

A study of the curves in a superimposed manner reveals that the tank circuits will pass only frequencies lying in the range of 925 kHz to 1075 kHz.

The "*S*" characteristic of the balanced-slope detector, shown in Fig. 7.12, represents this operation of the tank circuits. The "*S*" shape of the curve shows that the output voltage of the detector is linearly proportional to the input frequency within the tuning range of the tank circuits.

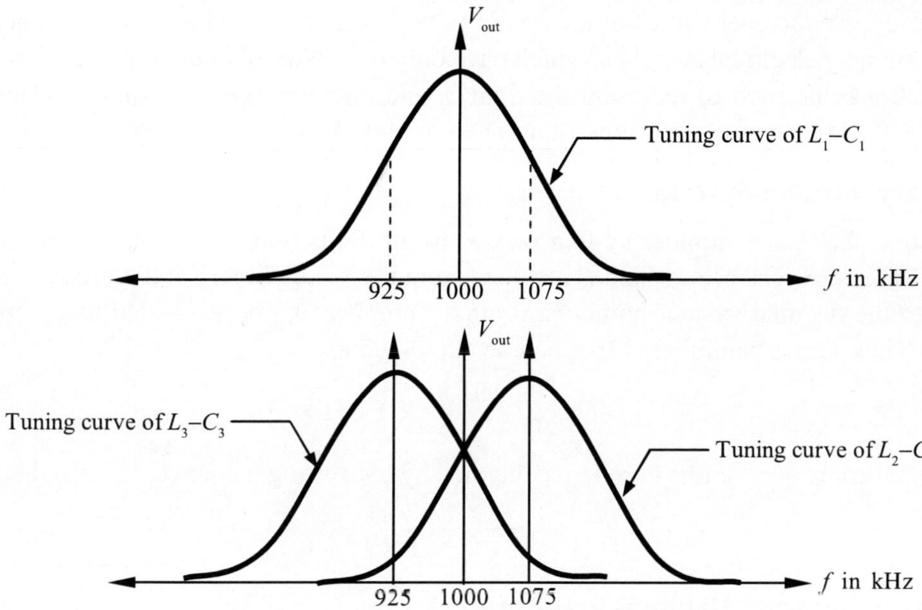

Fig. 7.11 Tuning characteristics of tank circuits.

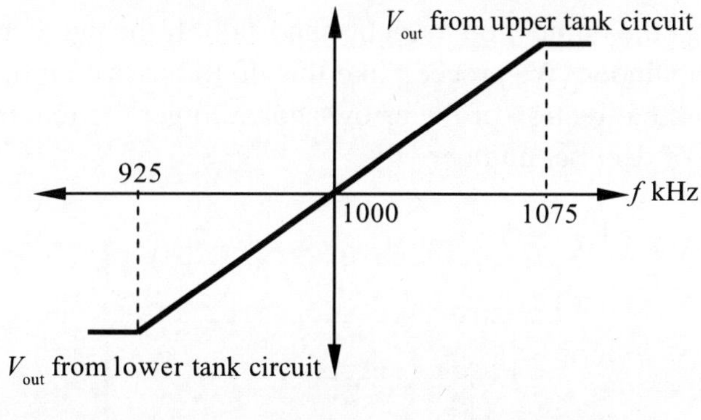

Fig. 7.12 "*S*" characteristics of balanced-slope detector.

7.2 NUMBER SYSTEM

7.2.1 Introduction

The decimal number system consisting of numbers 0, 1, 2, 3, 4, 5, 6, 7, 8, and 9, is the most widely used number system. This number system is said to have a base or radix of 10. Consider the decimal number 312. This number may be expressed as:

$$(312)_{10} = (3 \times 10^2) + (1 \times 10^1) + (2 \times 10^0)$$

where 3, 1, and 2 are the coefficients of the appropriate powers of ten. However, this is not suitable for representation by electrical switches, which have only two states of operation, viz., ON and OFF. An ON switch may be used to represent the digit 0, and an OFF signal the digit 7. This number system, where 0 and 1 are used to represent numbers is called the *binary number system.*

7.2.2 Binary Number System

As stated above, the binary number system makes use of digits 0 and 1 and has a radix of 2. The digits 0 and 1 are known as **bits**, which is a coined from the words **bi**nary and digi**ts**.

Similar to the decimal system numbers, a binary number can be expanded using appropriate powers of 2. Thus a binary number $(110)_2$ can be expressed as

$$(101)_2 = (1 \times 2^2) + (0 \times 2^1) + (1 \times 2^0)$$

The decimal equivalent of this number is obtained by converting the powers of 2 to their decimal values. Thus

$$(101)_2 = (1 \times 4) + 0 + (1 \times 1) = (5)_{10}$$

7.2.3 Conversion from Decimal to Binary.

Example 7.1 Convert $(19)_{10}$ into binary. The procedure used is illustrated in Fig. 7.13. In this, we divide the given number repeatedly by 2. In each division, if the number is exactly divisible by 2, we put a 0 as remainder against the number on its right-hand side. If the number is not exactly divisible by 2, we put a 1 as the remainder. We proceed like this up to the last digit, as shown in Fig. 7.13. We then read upwards, in the direction of the arrow shown to get the required binary number. We find the binary equivalent of decimal number.

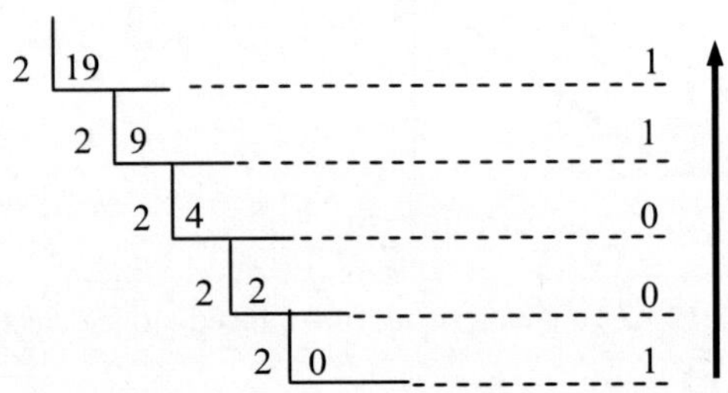

$$(19)_{10} = (10011)_2$$

Fig. 7.13 Decimal to binary conversion (method 1).

7.2.4 Decimal and Binary Fractions

Consider the decimal fraction 0.35. This may be expressed as

$$(0.35)_{10} = 3 \times 10^{-1} + 5 \times 10^{-2}$$

The numbers after the point symbol are expanded by using negative powers of 10, as shown. By using a similar method, we may express binary fractions. Consider, for example the binary number $(0.1101)_2$. This may be expanded as

$$(0.111)_2 = 1 \times 2^{-1} + 1 \times 2^{-2} + 1 \times 2^{-1}$$

The decimal equivalent of this binary fraction then becomes

$$(0.111)_2 = \frac{1}{2} + \frac{1}{4} + \frac{1}{8}$$
$$= \frac{7}{8}$$

The point symbol in the binary system may be called as the **binary point**.

7.2.5 Conversion from Decimal Fraction to Binary

To get binary fraction we multiply the given decimal fraction by 2. If the product of each number is greater than 1, we put a 1 below that number; otherwise we put a 0. This is illustrated in Table 7.1.

Example 7.2 Consider the decimal fraction 0.03125. To obtain its binary equivalent, we draw the table as shown in Table 7.1.

Table 7.1 Conversion from decimal-to-binary fraction

Number	0.03125	0.06250	0.1250	0.250	0.500
× 2	0.06250	0.1250	0.250	0.500	1.000
Remainder	0	0	0	0	1

From Table 7.1, we get the binary equivalent of decimal fraction $(0.03125)_{10}$ is $(0.00001)_2$.

Note: A binary fraction will converge if and only if the last number in the sequence in the decimal fraction is 0.5, so that $0.5 \times 2 = 1$; otherwise the fraction will continue.

7.2.6 Octal Number System

The basic numbers in the octal system are 0, 1, 2, 3, 4, 5, 6, and 7. The next number in this system after 7 is 10, and not 8. The subsequent numbers are 10, 11, 12, ... , 17, 20, 21, 21, ..., and so on.

7.2.7 Conversion of Decimal Number into Octal Number

To convert a given decimal number into octal number.

Example 7.3 To convert the given number $(7386)_{10}$ into octal, we proceed as shown in Table 7.2.

Table 7.2 Conversion from decimal into octal

Octal number	1	14	115	923	7386
Numerator ÷ 8	-	1	14	115	923
Decimal equivalent	1	6	3	3	2

The table is prepared from right to left. We first enter the decimal number to be converted in the rightmost top cell. Then divide it by 8. From Table 7.2, we obtain

$$(7386)_{10} \equiv (16332)_8$$

7.2.8 Conversion of Octal to Binary

Example 7.4 To convert octal number to binary, we first draw Table 7.3. Then we write below each number its binary equivalent in 3-bit form. We then get the entries as shown.

Table 7.3 Table for converting octal number to binary number

Octal number	1	6	5	4
Binary equivalent	001	110	101	100

7.2.9 Conversion of Binary to Octal

Example 7.5 We can apply a reverse process to yield the octal-equivalent of binary numbers. For this we draw Table 7.4 similar to Table 7.3.

As shown in the table, we divide the given binary bits into groups of 3 bits each, starting from the rightmost bit towards the left. Grouping continues until the first bit is reached. In this the first bit is 1, we have to add two 0's before it to make it into a group of 3 bits. The reading of the bottom row then gives the desired octal number.

Table 7.4 Table for converting binary number to octal number

Octal number	001	110	101	100
Binary equivalent	1	6	5	4

7.2.10 Hexadecimal Number System

Hexadecimal number system uses a base of 7. The initial 15 numbers of this system are 0, 1, 2, 3, 4, 5, 6, 7, 8, 9, A, B, C, D, E and F, respectively. In these numbers, A = 10, B = 11, C = 12, D = 13, E = 14, and F = 15. After F, the next numbers are 10, 11, 12, . . . , 19, 1A, 1B, . . . , and so on.

7.2.11 Convert Binary Numbers to Hexadecimal Numbers

Table 7.5 Table for converting binary number to hex number

Binary number	0001	1100	1010	1000
Hex equivalent	1	C	A	8

From Table 7.5, we obtain

$$(1110010101000)_2 \equiv (1CA8)_{16}$$

7.2.12 Convert Hexadecimal Numbers to Binary Numbers

Table 7.6 Table for converting Hex number to binary number

Binary number	1	C	A	8
Hex equivalent	0001	1100	1010	1000

From Table 7.6, we obtain

$$(1CA8)_{16} \equiv (1110010101000)_2$$

7.2.13 Binary Codes

Binary codes are used in communication and information systems for the following reasons:

- Codes are used to represent letters, numbers, and punctuation marks. For example, we may represent a letter in English alphabet by using a combination of binary digits (i.e., 1s and 0s).
- Codes may be used for secret transmission of information. Only persons with the prior knowledge of the codes will be able to decode them.
- Coding is required for maximum efficiency in signal transmission.
- By using multiplexing techniques, codes can be used for transmitting signals from different sources over the same channel. Great economy in signal transmission is achieved through such schemes.
- By using error-detecting codes, errors generated in signal transmission can be detected.
- Error-correcting codes are also available which can be used to correct errors generated during transmission.
- Self-complementing codes are used for arithmetic and logic operations in digital computers.
- Codes are also used for data compression by which large amounts of data are transmitted in very short duration of time.
- Binary codes are the major components in the synthesis (artificial generation) of speech and video signals.
- Codes are the basic building blocks of modern digital computer techniques such as picture and sound enhancement, picture morphing, etc.

7.2.14 Binary-Coded Decimal (BCD) Code

Decimal numbers can be represented directly by using binary digits in a coding scheme called the binary-coded decimal (BCD) code. This is illustrated by using an example.

Example 7.6 Consider decimal number 54. This is converted into BCD as shown below.

Decimal number	5	4	6
BCD code	0101	0100	0110

As shown above, the binary equivalent of 5 (using a 4-bit scheme) is 0101, that of 4 is 0100, and that of 6 is 0110. These are then combined to get the required BCD code as

$$(546)_{10} = (010101000110)_{BCD}$$

This method may be used to write any decimal number in its equivalent binary form.

Advantages of BCD code

- Any large decimal number can be easily converted into corresponding binary number.
- A person need remember only the binary equivalents of decimal numbers from 0 to 9.
- Conversion from BCD back to decimal is also very easy.

Disadvantages of BCD code

- The code is very inefficient; it requires several symbols to represent even small numbers.
- Binary addition and subtraction can lead to wrong answers.
- Special codes are required for arithmetic operations.
- This is not a self-complementing code.
- Conversion to other coding schemes requires special methods.

ONE-WORD/MULTIPLE-CHOICE QUESTIONS

M7.1 The height of a resonant antenna is given by ________.

M7.2 The expression for amplitude modulation is given by ________.

M7.3 The expression for frequency modulation is given by ________.

M7.4 The expression for phase modulation is given by ________.

M7.5 The expression for amplitude-modulation index is given by ________.

M7.6 The expression for frequency-modulation index is given by ________.

M7.7 The expression for phase-modulation-index is given by _______.

M7.8 The elementary radio receiver using a single diode is called a _______ receiver.

M7.9 A superheterodyne receiver converts all incoming frequencies to a constant frequency called the _______ frequency.

M7.10 There are _______ stages of IF amplifier in a transistor radio receiver.

M7.11 _______ modulator is the direct form of FM transmitter.

M7.12 _______ modulator is the indirect form of FM transmitter.

M7.13 Balanced-slope detector is a type of _______ detector.

M7.14 Balanced-slope detector suffers from the fact that it has _______ capacitors for tuning the circuit.

M7.15 Standard frequency deviation in the professional FM broadcast transmission is _______.

M7.16 Bit is a short form for _______.

M7.17 The number system that uses only the digits of 0 and 1 is known as the _______.

M7.18 The number system that uses the digits of 0 and 7 is known as the _______.

M7.19 The number system that uses the digits of 0 and 15 is known as the _______.

M7.20 The decimal equivalent of $(A)_{16}$ is _______.

M7.21 The binary equivalent of $(A)_{16}$ is _______.

M7.22 The binary equivalent of $(FFF)_{16}$ is _______.

M7.23 The octal equivalent of $(FFF)_{16}$ is _______.

M7.24 The BCD equivalent of $(100101110111)_2$ is _______.

M7.25 The octal equivalent of $(100)_{10}$ is _______.

REVIEW QUESTIONS

1. Give the general definition of modulation.
2. What is the need for modulation?
3. Derive the expression for amplitude modulation.

4. Define amplitude-modulation index.
5. Draw a neat sketch of an amplitude-modulation wave.
6. With suitable block schematics, explain the principle of operation of a low-power transmitter.
7. Explain with the help of a block diagram the operation of a high-power transmitter.
8. Compare low-power and high-power transmitters.
9. Describe the working of an isolation amplifier.
10. Describe the working of a master oscillator
11. What is a harmonic generator? Draw the circuit of a tuned amplifier and prove that it can act as a harmonic generator.
12. Explain the working of a simple crystal detector.
13. Explain with circuit diagram, the functioning of a crystal radio receiver.
14. What are the advantages and disadvantages of crystal radio receiver.
15. Using block diagram, explain the operation of a super heterodyne radio receiver.
16. Explain with relevant waveforms the principle of operation of frequency modulation.
17. Explain the principle of phase modulation.
18. What are the methods for the generation of frequency modulation?
19. Explain Armstrong modulation.
20. Explain the working of balanced-slope detector.

UNSOLVED PROBLEMS

P7.1 Compute the height of an antenna that can be used to transmit a frequency of 1 MHz. Assume that the antenna is a resonant one.

P7.2 Compute the height of an antenna that can be used to transmit a frequency of 1 GHz. Assume that the antenna is a resonant one.

P7.3 An amplitude modulation transmitter uses a carrier signal of $100\cos(2\pi \times 10^6 t)$ and a modulating signal of $70\cos(2\pi \times 10^4 t)$. Obtain the expression of the modulated wave.

P7.4 A frequency-modulation transmitter uses a carrier signal of $100\sin(2\pi \times 10^6 t)$ and a modulating signal of $70\sin(2\pi \times 10^4 t)$. Choose frequency deviation = 75 kHz. Obtain the expression of the modulated wave.

P7.5 A frequency-modulation transmitter uses a carrier signal of $100\sin(2\pi \times 10^6 t)$ and a modulating signal of $70\sin(2\pi \times 10^4 t)$. Assume proportionality constant $k_f = 1000$. Choose frequency deviation = 75 kHz. Obtain the expression of the modulated wave.

P7.6 Find the binary equivalent of decimal 100.

P7.7 Find the decimal equivalent of decimal $(0.101111)_2$.

P7.8 Find the binary equivalent of the decimal fraction 0.2315.

P7.9 Convert from decimal into octal given the decimal number is $(3312)_{10}$.

P7.10 Convert from octal into binary given that the octal number is $(3312)_8$.

P7.11 Convert from binary into octal given that the binary number is $(1101111001000011)_2$.

P7.12 Convert from binary into hex given that the binary number is $(1101111001000110)_2$.

P7.13 Convert from binary into hex given that the binary number is CCAB.

P7.14 Convert decimal number 269 into BCD.

P7.15 Convert decimal number 32198 into BCD.

P7.16 Convert BCD 1001 1000 0101 0001 0011 into decimal.

Boolean Algebra and Digital Logic

CHAPTER 8

8.1 INTRODUCTION

The algebra dealing with the binary numbers of 1 and 0 is known as **Boolean algebra**. This algebra was invented by George Boole in 1850. A Boolean variable x can assume only two values: 0 or 1. The binary system became prominent with the advent of electronic switches due to the fact that their ON and OFF states could be used to represent the binary numbers. There are three basic operations in Boolean algebra. They are the AND, OR, and NOT operations. Derived from these are the NAND (NOT AND) and NOR (NOT OR) operations, which are also considered as basic.

8.2 BASIC LOGIC FUNCTIONS

8.2.1 AND/NAND Functions

Consider the circuit shown in Fig. 8.1 (a). It consists of two switches A and B connected in series with a load resistance R_L to a DC power source $+V$. When both the switches are open, there will not be any current flow through R_L and the output voltage $V_o = 0$. The same is the situation when one of the switches is closed and the other opened. However, when both the switches are closed, current will flow through R_L and the output voltage $V_o = +V$.

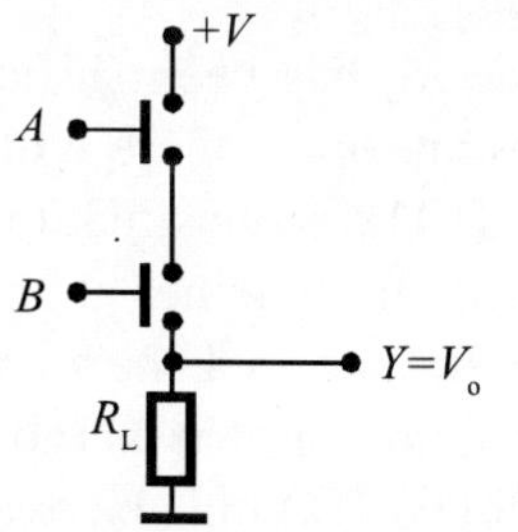

Fig. 8.1 (a) AND gate using switches.

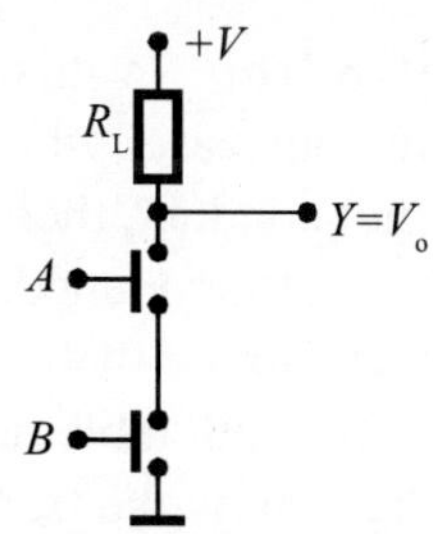

Fig. 8.1 (b) NAND gate using switches.

Let an open switch represent binary 0 and closed switch binary 1. Also, let the switches be designated as A and B, respectively, and the output as Y. Then we may express the switching situation shown in Fig. 8.1 (a) by the following statement

$$A = 0,\ B = 0,\ \text{then } Y = 0$$
$$A = 0,\ B = 1,\ \text{then } Y = 0$$
$$A = 1,\ B = 0,\ \text{then } Y = 0$$
$$A = 1,\ B = 1,\ \text{then } Y = 1$$

The above statements are usually expressed in a tabular form called the **truth table**. Table 8.1(a) shows the truth table representing these operations. The relations highlighted by the truth table given in Table 8.1(a) may be expressed by the equation:

$$Y = A \cdot B = AB \tag{8.1}$$

The operation defined by Eq. (8.1) is known as **AND** operation. AND operation may be represented by a dot or point symbol. It may also be represented without the dot. These are illustrated in Eq. (8.1). Equation (8.1) may also be stated as:

$$Y = A \text{ AND } B$$

"The output will be true if and only if both A AND B are true".

It can be seen that AND operation is in fact a *multiplication operation* in binary.

Table 8.1(a) AND

A	B	Y
0	0	0
1	0	0
0	1	0
1	1	1

Table 8.1(b) NAND

A	B	Y
0	0	1
0	1	1
1	0	1
1	1	0

Now consider Fig. 8.1(b). In this, the load resistance R_L has been shifted to the top of the switches. As in the previous case, when both the switches are open, there will be no current flow though R_L. If R_L does not conduct, then V_o will be $+V$ itself. The same situation exists if one of the switches is closed and the other opened. However, when both the switches are closed, current flows through R_L. However, we notice that both the closed switches act as a short across the output and hence V_o will be zero. The truth table describing this operation is shown in Table 8.1(b).

The relations highlighted by the truth table given in Table 8.1(a) may be expressed as

$$Y = \overline{AB} \tag{8.2}$$

The operation defined by Eq. (8.2) is known as NAND operation. Equation (8.2) may also be stated as

$$Y = \overline{A \text{ AND } B}$$

"The output will be true if and only if both *A* AND *B* are not true".

Figure 8.2(a) shows the symbol of an AND gate and 8.2(b) shows the symbol of a NAND gate. Notice the bubble at the output of the NAND symbol. In general, bubbles are used to denote negation (NOT) operation.

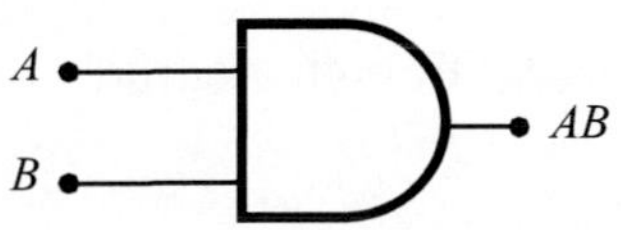

Fig. 8.2 (a) Symbol of AND gate.

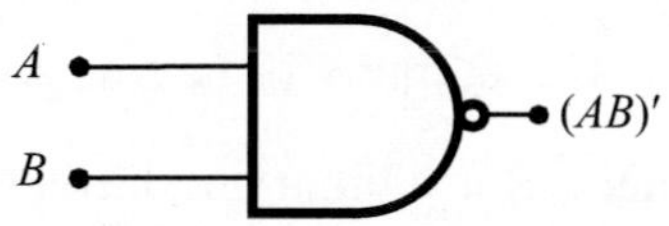

Fig. 8.2 (b) Symbol of NAND gate.

8.2.2 OR/NOR Functions

Consider the circuit shown in Fig. 8.3 (a). It consists of two switches *A* and *B* connected in parallel. A load resistance R_L and a DC power source $+V$ are connected in series with these switches. When both the switches are open, there will not be any current flow through R_L and the output voltage $V_o = 0$. However, if any one of the switches is closed or both the switches are closed, will flow through R_L and the output voltage $V_o = +V$. This operation is described in the truth table shown in Table 8.3(a).

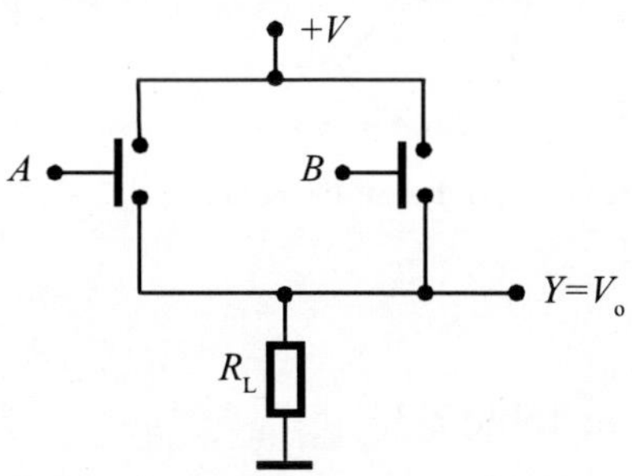

Fig. 8.3 (a) OR gate using switches.

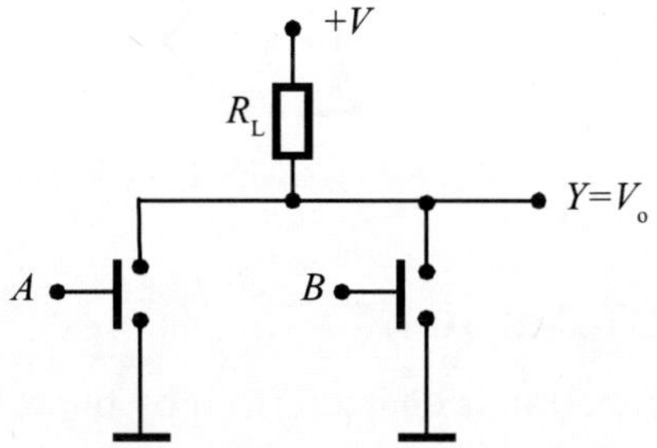

Fig. 8.3 (b) NOR gate using switches.

The relations highlighted by the truth table given in Table 8.3(a) may be expressed by the equation

$$Y = A + B \tag{8.3}$$

The operation defined by Eq. (8.3) is known as OR operation. In words, we can express the OR operation as:

$$Y = A \text{ OR } B$$

Table 8.2(a) OR

A	*B*	*Y*
0	0	0
0	1	1
1	0	1
1	1	1

Table 8.2(b) NOR

A	*B*	*Y*
0	0	1
0	1	0
1	0	0
1	1	0

The output will be true when "either *A* OR *B* OR both are true".

It can be seen that OR is a binary addition operation.

Now consider Fig. 8.3(b). In this, the load resistance R_L has been shifted to the top of the switches. As in the previous case, when both the switches are open, there will be no current flow though R_L. If R_L does not conduct, then V_o will be $+V$ itself. However, when one or both of the switches are closed current will flow through R_L and the output and hence V_o will be zero. The truth table describing this operation is shown in Table 8.2(b). The relations highlighted by the truth table given in Table 8.2(b) may be expressed as

$$y = \overline{A + B} \tag{8.4}$$

The operation defined by Eq. (8.4) is known as **NOR** operation. Figures 8.4(a) and (b) show, respectively, the symbols of OR and NOR gates.

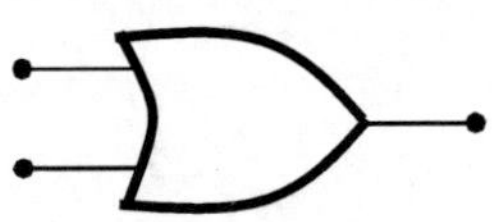

Fig. 8.4 (a) Symbol of OR gate.

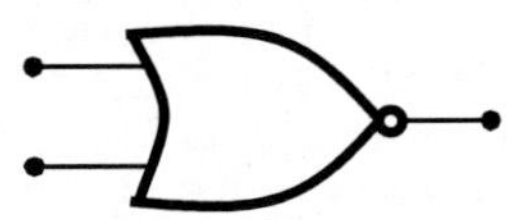

Fig. 8.4 (b) Symbol of NOR gate.

8.2.3 NOT Function

The NOT function is characterised by the relation shown in Table 8.3.

Table 8.3 NOT

A	$\overline{A}$
0	1
1	0

Figure 8.5 (a) shows a simple transistor amplifier configured as a NOT gate. The NOT gate is also known as **INVERTER**. When the input is zero, transistor is OFF and the output is V_{CC} (logic

1). When the input is a pulse of sufficient amplitude, the transistor turns on and the output drops to V_{CE} sat (= 0.2 V = logic 0). Thus the transistor performs NOT operation. Figure 8.5(b) shows the symbol of the NOT gate.

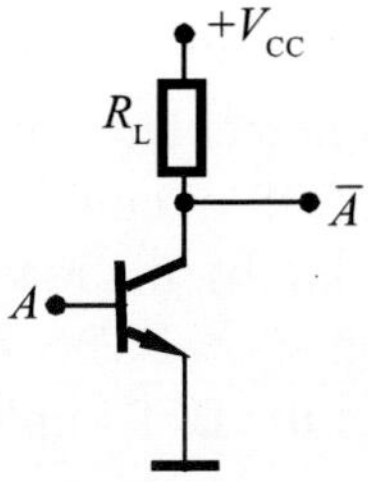

Fig. 8.5 (a) NOT gate using RTL.

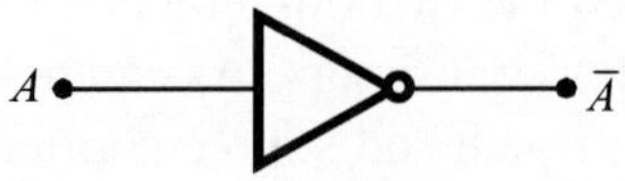

Fig. 8.5 (b) Symbol of NOT gate.

8.2.4 Implementation of AND/NAND Functions using Electronic Circuits

Resistor-Transistor Logic (RTL)

Figures 8.6(a) and (b), respectively show the implementation of AND and NAND operations using BJT's. In the scheme shown in Fig. 8.6(a), the switches of Fig. 8.1(a) are replaced by BJT switches. These circuits are known as the **Resistor-Transistor Logic (RTL)** AND and NAND gates, respectively.

Consider the RTL gate shown in Fig. 8.6(a). When both the transistors are OFF, the output across R_L = 0. The same situation exists even when one of the transistors is turned on. However, when both the transistors are turned-on, collector current flows through R_L to develop the output across it. This, then, is the AND operation. In a similar way, we can prove that the circuit of Fig. 8.6(b) performs NAND operation.

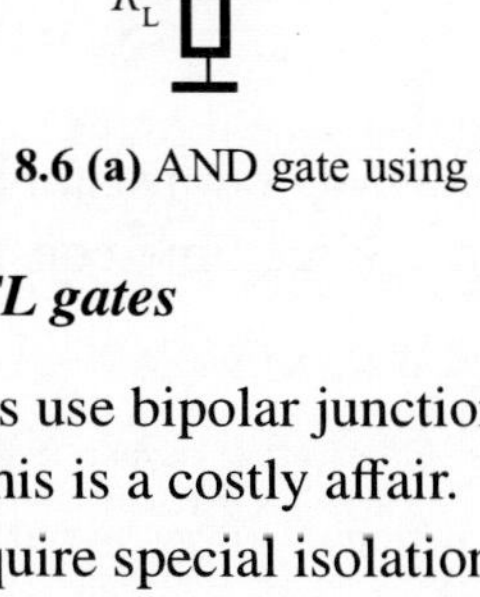

Fig. 8.6 (a) AND gate using BJT's.

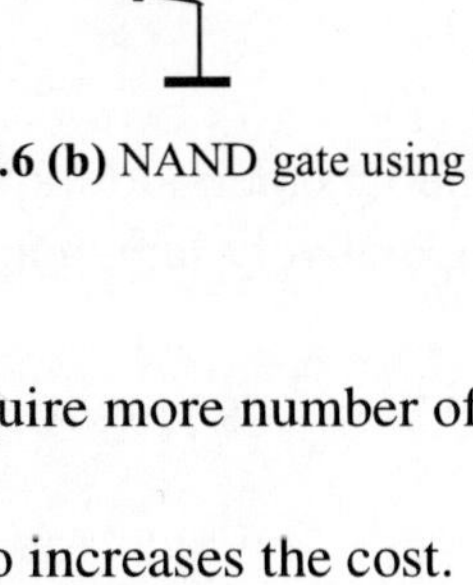

Fig. 8.6 (b) NAND gate using BJT's.

Demerits of RTL gates

- RTL gates use bipolar junction transistors, which require more number of steps for manufacturing. This is a costly affair.
- BJT's require special isolation techniques, which also increases the cost.

- They use resistors, which occupy large chip area. This also increases the cost.
- They are much less dense than MOS transistor logic IC gates.

NMOS Transistor Logic Gates

Figures 8.7(a) and (b) represent, respectively, the NMOS-transistor implementation of NAND and NOR functions. These circuits are similar to their counterparts implemented using BJT. The major difference between the two implementation schemes is that in NMOS circuits all components are formed from NMOS transistors. As can be seen from Figs. 8.7(a) and (b), the load resistance is also an NMOS, with its gate and source shorted. This saves costly chip space. The working principles of these circuits are exactly similar to those described in the case of the BJT implementations.

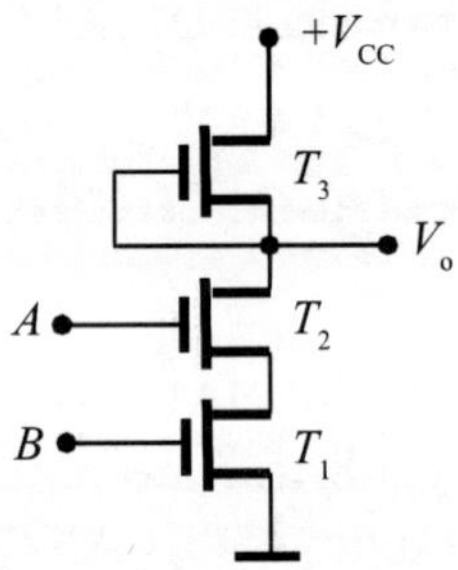

Fig. 8.7 (a) NAND gate using MOSFET's.

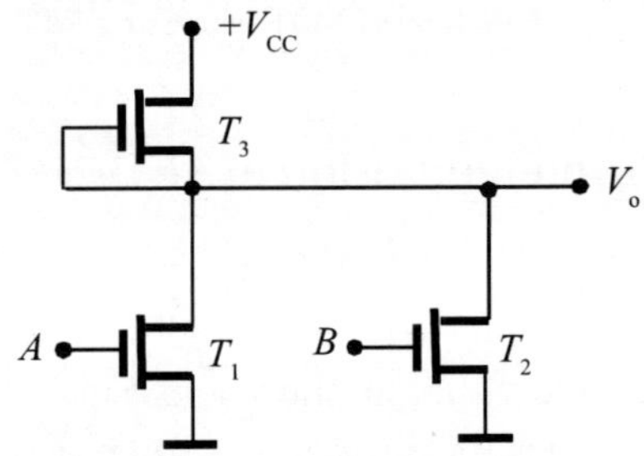

Fig. 8.7 (b) NOR gate using MOSFET's.

Merits of NMOS Logic gates

- NMOS gates use NMOS transistors, which require lesser number of steps for manufacturing. Thus cost is reduced.
- No special isolation is required, which also reduces the cost of these gates.
- All components are formed from the basic NMOS transistor; hence chip space saved.
- NMOS transistor logic IC gates have maximum packing density.
- Very small power supply requirement.
- Very low loading effects.
- Can drive several other logic gates.
- Large noise margin.

Demerits

- The major demerit is their slow speed owing to the fact that NMOS transistors are also capacitors with comparatively large capacitance values and hence large time-constants associated with them.
- Power dissipation is much higher than CMOS logic gates.

Diode Logic (DL) Gates

Figure 8.8 (a) shows a diode AND gate. When the inputs $A = B = 0$, the cathodes of the diodes are grounded and the out put is 0. The same is the condition, even when one of the inputs is 0 and the other is 1. When both $A = B = 1$, the input terminals are open and no current flows through R_L. Then the output $Y = +V = 1$. These actions indicate that the circuit shown in Fig. 8.8(a) performs AND operation.

Figure 8.8(b) shows a diode OR gate. When the inputs $A = B = 0$, the inputs to the diodes are 0 and the out put is 0. But when one of the inputs or both the inputs are 1, current flows through R_L. Then the output $Y = 1$. These indicate that the circuit shown in Fig. 8.8.b performs OR operation.

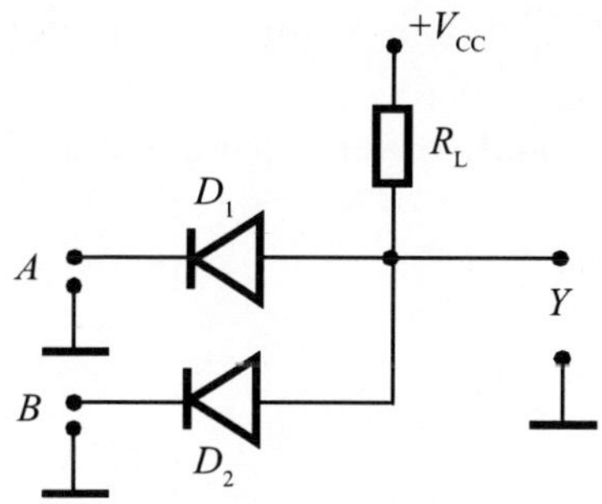

Fig. 8.8 (a) Diode AND gate.

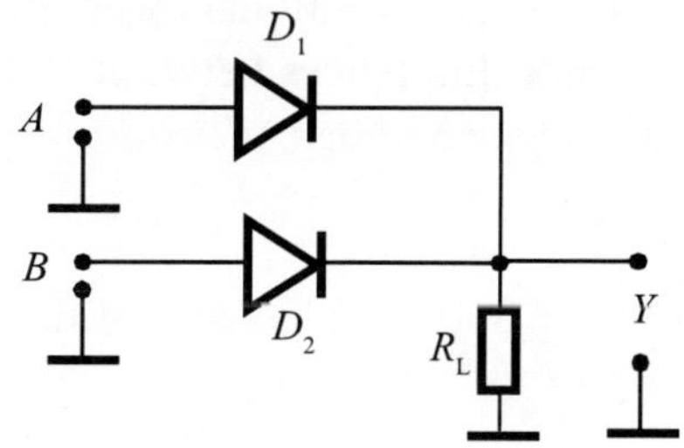

Fig. 8.8 (b) Diode OR gate.

Diode Transistor Logic (DTL) Gates

It can be easily verified that all logic functions can be generated if NAND or NOR gates are available. This is because these negative logic gates can perform both negative and positive operations (see Section 8.3 - Demorgan's Laws).

Diode logic can be converted into NAND or NOR gates by adding transistor inverters to them. Figure 8.9 shows a Diode-Transistor Logic (DTL) NAND gate.

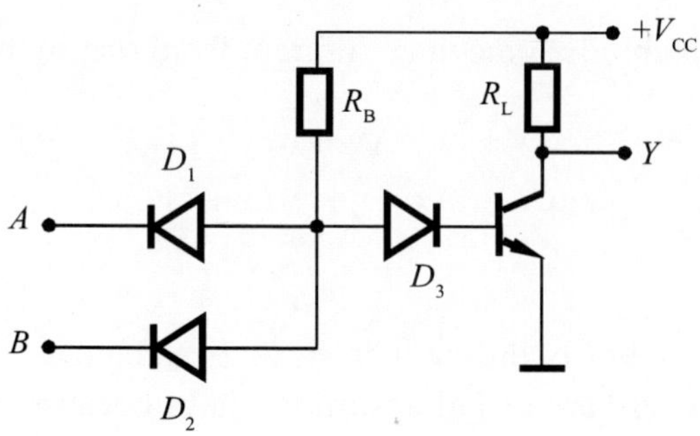

Fig. 8.9 Diode-Transistor Logic (DTL) AND gate.

The diode section consisting of D_1 and D_2 performs AND operation. The transistor inverts this to yield NAND operation at its output. Diode D_3 ensures that the transistor turns-on only if the output of the diode gate is greater than 1.6 V. This prevents noise voltages from turning-on the gate.

DTL gates have been replaced by TTL (Transistor-Transistor Logic) gates in which all the devices such as diodes and resistors are also made by appropriately converting transistors into such components.

8.3 LAWS OF BOOLEAN ALGEBRA

As stated earlier, Boolean algebra is a special category of conventional algebra whose variables can assume only two values: 0 and 1. Hence the rules of this algebra have some similarities and some dissimilarities from the conventional algebra.

One of the major differences between Boolean algebra and conventional algebra. Some of the Boolean laws are postulated below.

Commutative Law

Commutative law in Boolean algebra is similar to that in the regular law.

$$x + y = y + x$$
$$x \cdot y = y \cdot x$$

Associative Law

Associative law in Boolean algebra is similar to that in the regular law.

$$x + (y + z) = (x + y) + z$$
$$x\,(yz) = (xy)\,z$$

Distributive Law

Distributive law in Boolean algebra is somewhat different from that in the regular law.

$$x\,(y + z) = xy + xz$$
$$x + (yz) = (x + y)\,(x + z)$$

Absorption Laws

Absorption law in Boolean algebra is different from that in regular law; in fact such a law can not exist in regular algebra. These are called absorption laws because one of the variables is seen absorbed in the final results.

1. $x + xy = x$, $x\,(x + y) = x$
2. $x + x'y = x + y$, $x\,(x' + y) = xy$
3. $x' + xy = x' + y$, $x'\,(x + y) = x'y$

Idempotency

$$x + x = x$$

$$xx = x$$

Consensus theorem

$$xy + yz + x'z = xy + x'z$$

$$(x + y)\ (y + z)\ (x' + z) = (x + y)\ (x' + z)$$

Some other basic laws

1.	$x + 1 = 1$	2.	$x + 0 = x$	3.	$x + x' = 1$	4.	$1 + 1 = 1$	5.	$1 + 0 = 1$
	$x \cdot 1 = x$		$x \cdot 0 = 0$		$x \cdot x' = 0$		$1 \cdot 1 = 1$		$1 \cdot 0 = 0$

Demorgan's Laws

Demorgan's laws are one of the most important set of Boolean algebraic laws. Such a set of laws are not available in conventional algebra. These laws are useful for converting NAND gates to NOR gates and vice versa.

$$(x + y)' = x'y'$$

$$(xy)' = x' + y'$$

8.4 TRUTH TABLES

Boolean algebraic expressions can be tested for their validity by using truth tables. Tables 8.1 to 8.3 represent truth tables associated with various logic functions.

Example: DeMorgan's Law

Prove that $(x + y)' = x'y'$

Solution

To prove the given DeMorgan's law, we draw the truth table as shown in Table 8.4(a). The entries in the table are made by considering the components of the given equation in individual columns. We find that the entries in columns 4 and 7 are the same, which is the desired proof.

Table 8.4(a)

x	y	$x + y$	$(x + y)'$	x'	y'	$(xy)'$
0	0	0	1	1	1	1
0	1	1	0	1	0	0
1	0	1	0	0	1	0
1	1	1	0	0	0	0

Example: DeMorgan' Law

Prove that $(xy)' = x' + y'$

Solution

To prove the given DeMorgan's law, we draw the truth table as shown in Table 8.4(b). As before, the entries in the table are made by considering the components of the given equation in individual columns. We find that the entries in columns 4 and 7 are the same, which is the desired proof.

Table 8.4(b)

x	y	xy	$(xy)'$	x'	y'	$x' + y'$
0	0	0	1	1	1	1
0	1	0	1	1	0	1
1	0	0	1	0	1	1
1	1	1	0	0	0	0

8.5 DUALITY OF EQUATIONS

It can be seen that every law described in the previous section has two forms: one expressed as the **sum of products** (SOP) of variables and the other the **product of sum** (POS) of variables. Both these forms are a unique feature of the Boolean algebra and are equally valid. These equations are called **dual** equations and the property feature is called **duality**.

In Boolean duality, to get the dual relation, we change the AND functions to OR functions and vice versa. For example, consider the distributive law:

$$x(y + z) = xy + xz$$

Using the principle given above, the dual of this equation can be obtained by interchanging AND's and OR's. Thus dual of $x(y + z) = xy + xz$ is

$$x + (yz) = (x + y)(x + z)$$

Example: Consensus Theorem

$$xy + yz + x'z = xy + x'z$$

Solution

By comparing the entries given in columns 8 and 9 of Table 8.5, we find that they are one and the same, thus proving the validity of the consensus theorem.

Thus we conclude that Boolean functions can be tested for their validity using appropriate truth tables.

Table 8.5

x	y	z	xy	x'	$x'z$	yz	s_1	s_2
0	0	0	0	1	0	0	0	0
0	0	1	0	1	1	0	1	1
0	1	0	0	1	0	0	0	0
0	1	1	0	1	1	0	1	1
1	0	0	0	0	0	0	0	0
1	0	1	1	0	0	0	1	1
1	1	0	1	0	0	0	1	1
1	1	1	1	0	0	1	1	1

8.6 IMPLEMENTATION OF LOGIC FUNCTIONS USING LOGIC GATES

Example: Implement using basic logical gates the function

$$Y = (x + y)(x + z)$$

Solution

The implementation is shown in Fig. 8.10.

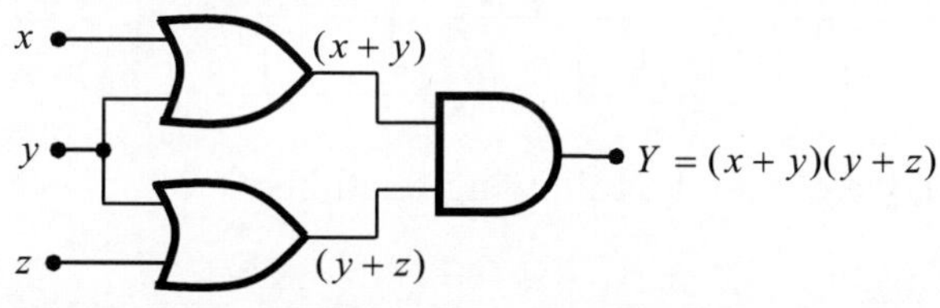

Fig. 8.10 Implementation of the expression $Y = (x + y)(y + z)$.

Example: Write the Boolean expression for the output of the circuit shown in Fig. 8.11.

Solution

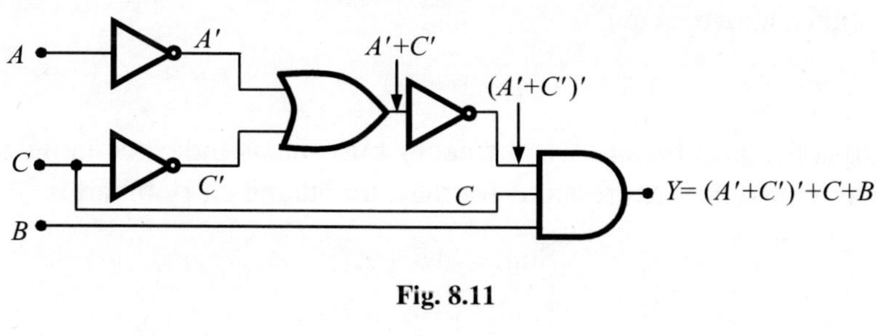

Fig. 8.11

From the circuit diagram shown in Fig. 8.11, we get the output as

$$Y = (A' + C')' + C + B$$

Now, using DeMorgan's laws, the term

$$(A' + C')' = [(A') + (C')]' = (A')'(C')' = AC$$

Then

$$Y = AC + C + B = C(A + 1)B = BC$$

8.7 BINARY ADDERS

8.7.1 Half-Adder

The basic binary adder is used to add just two bits. In the least-significant bit position, for addition, there will be no carry. The binary adder that adds two bits is called a half-adder. Table 8.6 shows the truth table of a half-adder.

Table 8.6 Truth table of half-adder

A	*B*	**Sum**	**Carry**
0	0	0	0
0	1	1	0
1	0	1	0
1	1	0	1

From columns *A* and *B*, we obtain the following relations:

$$0 + 0 = 0$$
$$0 + 1 = 1$$
$$1 + 0 = 1$$
$$1 + 1 = 0$$

In the last row shown above, actually

$$1 + 1 = 10$$

Of the sum 10, 0 is the **sum** bit and 1 is the **carry** bit. This is indicated in the truth table as the sum bit and carry bit. The logical expressions for the sum bit and carry bit are:

$$\text{Sum} = \overline{A}B + A\overline{B}$$
$$\text{Carry} = AB$$

The half-adder can be implemented using the set-up shown in Fig. 8.12. The sum and carry are shown as separate outputs. The summing circuit is known as **EXOR (Exclusive OR)** gate, and is available as an IC. Figure 8.13 shows the symbol of the EXOR (XOR) gate.

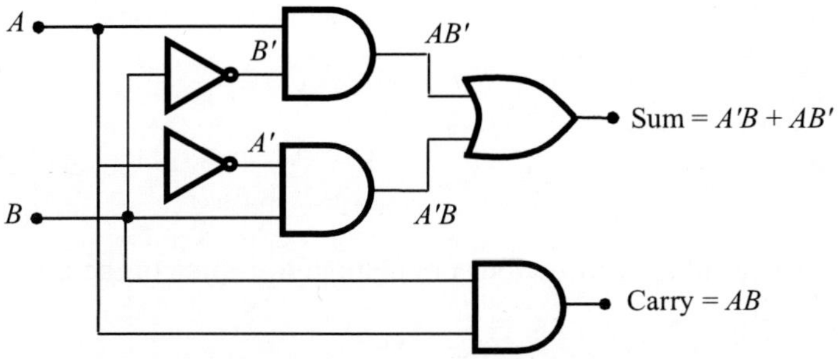

Fig. 8.12 Implementation of half-adder.

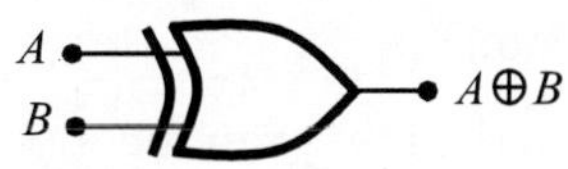

Fig. 8.13 Symbol of XOR gate.

8.7.2 Full-Adder

The full-adder is used to add three bits, the third bit being the carry bit. From the second bit position onwards, for addition carry bit also has to be added. The binary adder that adds three bits is called a full-adder. Table 8.7 shows the truth table of a full-adder.

Table 8.7

x	y	z	xy	x'	$x'z$	yz	s_1	s_2
0	0	0	0	1	0	0	0	0
0	0	1	0	1	1	0	1	1
0	1	0	0	1	0	0	0	0
0	1	1	0	1	1	0	1	1
1	0	0	0	0	0	0	0	0
1	0	1	1	0	0	0	1	1
1	1	0	1	0	0	0	1	1
1	1	1	1	0	0	1	1	1

Form Table 8.7, we obtain the sum

$$S = A\overline{BC} + \overline{A}B\overline{C} + \overline{AB}C + ABC$$

and the carry

$$C = AB\overline{C} + A\overline{B}C + \overline{A}BC + ABC$$

8.8 UNIVERSAL GATES

Any logic function can be constructed using NAND gates or NOR gates only. Hence they are called universal gates.

Example: Obtain an all NAND implementation of the logic function

$$Y = A + B$$

Solution

First write the given function in the form. The double barring does not change the value of the given function. Thus

$$\overline{\overline{Y}} = \overline{\overline{A + B}}$$

Now, retain the topmost bar and apply DeMorgan's law to the bottom bar. This yields the expression

$$Y = \overline{\overline{A}\,\overline{B}}$$

This can be implemented using the circuit shown in Fig. 8.14.

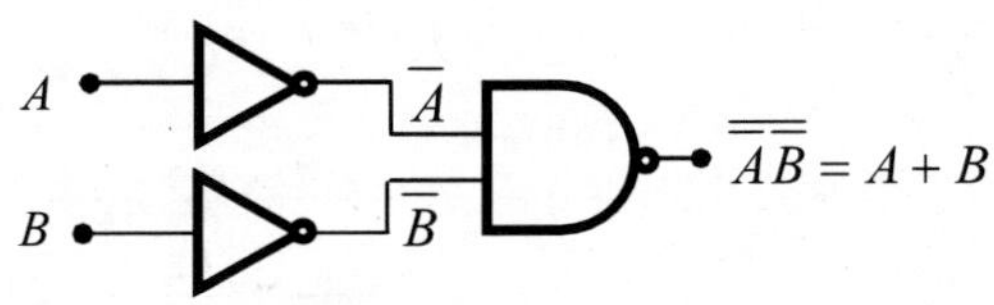

Fig. 8.14 All NAND implementation OR.

ONE-WORD/MULTIPLE-CHOICE QUESTIONS

M8.1 The algebra dealing with the binary number system is known as the ________ algebra.

M8.2 The three basic logic functions are ________ and ________.

M8.3 Two switches and a load connected in series will form ________ logic operation.

M8.4 Two switches connected in parallel and a load connected in series with them will form ________ logic operation.

M8.5 Two open switches connected in parallel and two closed switches connected in series with the parallel combination will form ________ logic operation.

M8.6 Two open switches connected in series and two closed switches connected in parallel with the series combination will form ________ logic operation.

M8.7 A CMOS logic circuit dissipates ________ amount of power.

M8.8 A CMOS logic circuit dissipates negligible amount of power since it is _______ at most instants of time.

M8.9 NAND and NOR gates are called _______ gates.

M8.10 _______ and _______ are universal gates.

M8.11 TTL stands for _______.

M8.12 ECL stands for _______.

M8.13 RTL stands for _______.

M8.14 DTL stands for _______.

M8.15 I^2L stands for _______.

M8.16 $x + x = x$ is an example for _______ binary law.

M8.17 $xx = x$ is an example for _______ binary law.

M8.18 $x + x'y = x + y$ is an example for _______ binary law.

M8.19 $x + xy = x$ is an example for _______ binary law.

M8.20 $xy + yz + x'z = xy + x'z$ is an example for _______ binary law.

REVIEW QUESTIONS

1. What are basic logic functions?
2. Illustrate how logic functions AND and NAND can be realised using switches.
3. Explain how OR and NOR functions can be realised using switches.
4. Explain the RTL family with its merits and demerits.
5. Discuss the NMOS family with its merits and demerits.
6. Explain the DTL family with its merits and demerits.
7. Explain TTL family with its merits and demerits.
8. Give a few laws of Boolean Algebra.
9. State and prove DeMorgan's Laws.
10. Explain duality equations.
11. Implement a few basic logic functions using logic gates.
12. Explain with logic diagram a half-adder.

13. Explain a full-adder circuit.

14. What are universal gates?

UNSOLVED PROBLEMS

P8.1 Give the name of the law applicable in the following case.

1. $a + b = b + a$
   $ab = ba$
2. $a + (b + c) = (a + b) + c$
   $a\,(bc) = (ab)\,c$
3. $w\,(x + y + z) = wx + wy + wz$
4. $a + abc = a$
   $a\,(a + b + c) = a$
5. $a + a'b = a + b$
   $a\,(a' + b) = ab$
6. $a' + ab = a' + b$
   $a'\,(a + b) = a'b$
7. $a + a = a$
   $aa = a$

P8.2 Prove: A full adder = Sum of two half-adders

P8.3 Obtain an all NAND implementation of the logic function $Y = A + B + C$.

P8.4 Prove using truth table the Commutative Law.

Solutions

Chapter-1

One-Word Multiple-Choice Questions

M1.1 Cat's whisker

M1.2 0.3 V

M1.3 0.3 V

M1.4 High

M1.5 Mobile

M1.6 False

M1.7 Ohm's

M1.8 nev_{d} or $ne\mu\epsilon$

M1.9 $ne\mu$

M1.10 Ohm's

M1.11 $eD_n\dfrac{dn}{dx}$

M1.12 Impact

M1.13 Field

M1.14 Heavily

M1.15 Medium/heavily

M1.16 Temperature

M1.17 $-eD_p\,(dp/dx)$

M1.18 V_{m}/π

M1.19 $2V_{\mathrm{m}}/\pi$

M1.20 $2V_{\mathrm{m}}/\pi$

M1.21 $V_{\mathrm{m}}/\sqrt{2}$

M1.22 $V_{\mathrm{m}}/\sqrt{2}$

M1.23 $V_{\mathrm{m}}/2$

M1.24 81.2%

M1.25 81.2%

M1.26 40.6%

M1.27 0.48

M1.28 0.48

M1.29 1.21

M1.30 $2V_{\mathrm{m}}$

M1.31 V_{m}

M1.32 V_{m}

Unsolved Problems

P1.1 Using the relation $k_j = W_j/\sqrt{V_j}$, we find $k_j = 6.45$ μm/V$^{1/2}$ **(Ans)**

P1.2 Using $k_j = W_j/\sqrt{V_j}$, we get $k_j = 5.16$ μm/V$^{1/2}$ **(Ans)**

P1.3 Using the relation $k_j = W_j/V_j^{1/3}$, we get $k_j = 2.99$ μm/V$^{1/3}$ **(Ans)**

P1.4 We have $k_j = W_j/V_j^{1/3} = 3.56$ μm/V$^{1/3}$ **(Ans)**

P1.5 Using the relation $V_j^{1/3} = \dfrac{W_j}{k_j} = \dfrac{3.56}{4} = 0.89$, we find the junction potential as $V_j = (0.89)^3 = 0.7$ V **(Ans)**

P1.6 Compute the value of electron drift current density J_{ndrift} in a given semiconductor material, if the number of electrons in one cm^3 of that material is $2 \times 2 \times 10^{16}$. The mobility of electrons in that material is 1000 cm^2/V-s, and the applied electric field intensity is 15 V/cm.

P1.7 Using the relation given below, we obtain

$$J_{ndrift} = ne\mu_n E = 2 \times 10^{16} \times 1.6 \times 10^{-19} \times 1000 \times 15 = 48 \text{ A/cm}^2 \textbf{ (Ans)}$$

P1.8 Using the relation given below, we obtain

$$J_{ndrift} = ne\mu_n E = 2 \times 10^{22} \times 1.6 \times 10^{-19} \times 100 \times 150 = 48 \times 10^6 \text{ A/m}^2 \textbf{ (Ans)}$$

P1.9 Substituting in the expression, we obtain $\sigma = ne\mu_n = 1 \times 10^{16} \times 1.6 \times 10^{-19} \times 1300 = 2.08$ S/cm **(Ans)**

P1.10 Using the relation $\rho = 1/\sigma = 1/ne\mu_n = 0.481$ Ω-cm **(Ans)**

P1.11 Using the relation $\mu_n = \dfrac{J_{ndrift}}{neE} = \dfrac{50}{2 \times 10^{16} \times 1.6 \times 10^{-19} \times 15} = 1041.67$ cm^2/V-s **(Ans)**

P1.12 (a) $I_{max} = \dfrac{V_{max}}{r_d + r_t + R_L} = \dfrac{12\sqrt{2}}{1034(= 12 + 22 + 1000)} = 16.413$ mA

$$I_D = \frac{I_{max}}{\pi} = \frac{16.413}{\pi} = 5.224 \text{ mA } \textbf{(Ans)}$$

Hence the DC output voltage

$$V_D = I_D R_L = 5.224 \text{ mA} \times 1000 = 5.224 \text{ V } \textbf{(Ans)}$$

(b) $I_r = \dfrac{I_{max}}{2} = \dfrac{16.413}{2} = 8.206$ V **(Ans)**

Hence the rms output voltage

$V_r = I_r R_L = 8.206 \times 1 = 8.206$ V **(Ans)**

(c) $\gamma = \sqrt{\dfrac{I_r^2 - I_D^2}{I_D^2}} = \sqrt{\dfrac{8.206^2 - 5.224^2}{5.224^2}} = 1.21$ **(Ans)**

(d) $\eta = \dfrac{I_D^2 R_L}{I_r^2 (r_t + r_d + R_L)} = \dfrac{(5.224^2 \times 10^{-6}) \times (1 \times 10^3)}{8.2061^2 \times 10^{-6} \times (1034)} = 39.2\%$ **(Ans)**

P1.13 (a) $I_{max} = \dfrac{V_{max}}{R_L + r_t + r_d} = \dfrac{12\sqrt{2}}{1034} = 16.413$ mA

$I_D = \dfrac{2I_{max}}{\pi} = \dfrac{16.413 \times 2}{\pi} = 10.45$ mA **(Ans)**

$V_D = I_D R_L = 10.45 \text{ mA} \times 1000\ \Omega = 10.45$V **(Ans)**

(b) $I_r = \dfrac{I_{max}}{\sqrt{2}} = \dfrac{16.413}{\sqrt{2}} = 11.606$ mA **(Ans)**

$V_r = I_r R_L = 11.606 \text{ mA} \times 1000 = 11.606$ V **(Ans)**

(c) $\gamma = \sqrt{\dfrac{I_r^2 - I_D^2}{I_D^2}} = \sqrt{\dfrac{11.606^2 - 10.45^2}{10.45^2}} = 0.483$ **(Ans)**

(d) $\eta = \dfrac{I_D^2 R_L}{I_r^2 (r_t + r_d + R_L)} = \dfrac{(10.45^2 \times 10^{-6}) \times (1 \times 10^3)}{11.606^2 \times 10^{-6} \times (1034)} = 78.41\%$ **(Ans)**

P1.14 (a) For the full-wave bridge connection, we find that a 12-0-12-V transformer can actually deliver 24-V rms AC since the center tap is not useful in this case.

$$I_{max} = \frac{24\sqrt{2}}{1034} = 32.83 \text{ mA}$$

and hence

$$I_D = \frac{2I_{max}}{\pi} = \frac{2 \times 32.83 \text{ mA}}{\pi} = 20.9 \text{ mA } \textbf{(Ans)}$$

$$V_D = I_D R_L = 20.9 \text{ mA} \times 1000 = 20.9 \text{ V } \textbf{(Ans)}$$

(b) $I_r = \dfrac{I_{max}}{\sqrt{2}} = \dfrac{32.83 \text{ mA}}{\sqrt{2}} = 23.21$ mA **(Ans)**

$V_r = I_r R_L = 23.21 \text{ mA} \times 1000 = 23.21$ V **(Ans)**

(c) $\gamma = \sqrt{\dfrac{I_{\rm r}^2 - I_{\rm D}^2}{I_{\rm D}^2}} = \sqrt{\dfrac{23.21^2 - 20.9^2}{20.9^2}} = 0.483$ **(Ans)**

(d) $\eta = \dfrac{I_{\rm D}^2 R_L}{I_{\rm r}^2 (r_{\rm t} + r_{\rm d} + R_{\rm L})} = \dfrac{20.9^2 \times 10^{-6} \times \left(1 \times 10^3\right)}{23.21^2 \times 10^{-6} \times (1034)} = 78.42\%$ **(Ans)**

(i) Centre-tapped rectifier

(a) $I_{\max} = \dfrac{V_{\max}}{R_{\rm L} + r_{\rm t} + r_{\rm d}} = \dfrac{9\sqrt{2}}{2260} = 5.632$ mA

$I_{\rm D} = \dfrac{2I_{\max}}{\pi} = \dfrac{2 \times 5.632 \times 10^{-3}}{\pi} = 3.59$ mA **(Ans)**

$V_{\rm D} = I_D R_L = (3.59 \text{ mA}) \times 2200 = 7.9$ V **(Ans)**

(b) $I_{\rm r} = \dfrac{I_{\max}}{\sqrt{2}} = \dfrac{5.632 \text{ mA}}{\sqrt{2}} = 3.982$ mA **(Ans)**

$V_{\rm r} = I_r R_L = (3.982 \text{ mA}) \times 2200 = 8.76$ V **(Ans)**

(c) $\gamma = \sqrt{\dfrac{I_{\rm r}^2 - I_{\rm D}^2}{I_{\rm D}^2}} = \sqrt{\dfrac{3.982^2 - 3.59^2}{3.59^2}} = 0.48$ **(Ans)**

(d) $\eta = \dfrac{I_{\rm D}^2 R_L}{I_{\rm r}^2 (r_{\rm t} + r_{\rm d} + R_{\rm L})} = \dfrac{\left(3.59^2 \times 10^{-6}\right) \times \left(2.2 \times 10^3\right)}{3.982^2 \times 10^{-6} \times (2260)} = 79.12\%$ **(Ans)**

(e) PIV $= 2V_{\max} = 2 \times 9 \times \sqrt{2} = 25.46$ V **(Ans)**

(ii) Bridge rectifier

(a) For the full-wave bridge connection, we find that a 9-0-9-V transformer can actually deliver 18-V rms AC since the center tap is not used in this case.

$$I_{\max} = \frac{18\sqrt{2}}{2260} = 11.264 \text{ mA}$$

and hence

$$I_{\rm D} = \frac{2I_{\max}}{\pi} = \frac{2 \times 11.264 \text{ mA}}{\pi} = 7.171 \text{ mA } \textbf{(Ans)}$$

$$V_{\rm D} = I_{\rm D} R_{\rm L} = (7.171 \text{ mA}) \times 2200 = 15.78 \text{ V } \textbf{(Ans)}$$

(b) $I_{\rm r} = \dfrac{I_{\max}}{\sqrt{2}} = \dfrac{11.264 \text{ mA}}{\sqrt{2}} = 7.965$ mA **(Ans)**

$V_{\rm r} = I_r R_L = (7.965 \text{ mA}) \times 2200 = 17.52$ V **(Ans)**

(c) $\gamma = \sqrt{\dfrac{I_r^2 - I_D^2}{I_D^2}} = \sqrt{\dfrac{7.965^2 - 7.171^2}{7.171^2}} = 0.48$ **(Ans)**

(d) $\eta = \dfrac{I_D^2 R_L}{I_r^2 (r_t + r_d + R_L)} = \dfrac{7.171^2 \times 10^{-6} \times \left(2.2 \times 10^3\right)}{7.965^2 \times 10^{-6} \times (2260)} = 78.90\%$ **(Ans)**

(e) Since the peak voltage of $(2 \times 9) \times \sqrt{2} = 25.46$ V is shared by two diodes, the peak-inverse voltage (PIV) of each diode is given by

$$\text{PIV} = V_{\text{max}} = 9 \times \sqrt{2} = 12.73 \text{ V}$$

(f) Results can be compared from the values obtained in each of the above cases. These are compared in the following table.

Parameter	**Centre-tapped**	**Bridge**
I_D (DC output current)	3.59 mA	7.171 mA
V_D (DC output voltage)	7.9 V	15.8 V
I_r (rms output current)	3.982 mA	7.965 mA
V_r ((rms output voltage)	8.76 V	17.52 V
γ (ripple factor)	0.48	0.48
η (efficiency)	79.12%	78.9%
PIV	25.46 V	12.73 V

P1.15 (i) Centre-tapped mode

(a) $I_{\text{max}} = \dfrac{V_{\text{max}}}{r_t + r_d + R_L} = \dfrac{15\sqrt{2}}{1 + 5 + 100} = 200$ mA

$I_D = \dfrac{2 \times 200 \text{ mA}}{\pi} = 127.32$ mA **(Ans)**

$V_D = 217.32 \times 100 = 12.732$ V **(Ans)**

(b) $I_r = \dfrac{I_{\text{max}}}{\sqrt{2}} = \dfrac{200 \text{ mA}}{\sqrt{2}} = 141.2$ mA

$V_r = 141.2 \times 0.1 = 14.12$ V **(Ans)**

(c) $\gamma = \sqrt{\dfrac{V_r^2 - V_D^2}{V_D^2}} = \sqrt{\dfrac{14.12^2 - 12.73^2}{12.73^2}} = 0.48$ **(Ans)**

(d) $\eta = \dfrac{127.32^2 \times 100}{141.2^2 \times 106} = 76.7\%$ **(Ans)**

(e) PIV = $2 \times 15\sqrt{2} = 42.43$ V **(Ans)**

(f) Rating of the load resistance $= I_D^2 R_L = 127.32^2 \times 0.100 = 1.62$ W **(Ans)**

Usually, for safety, we choose R_L having a rating of $2 \times 1.62 = 3.24$ W **(Ans)**.

(ii) Bridge mode

(a) $I_{max} = \dfrac{V_{max}}{r_t + r_d + R_L} = \dfrac{2 \times 15\sqrt{2}}{1 + 5 + 100} = 400$ mA

$I_D = \dfrac{2 \times 400 \text{ mA}}{\pi} = 254.81$ mA **(Ans)**

$V_D = 254.81 \times 100 = 25.48$ V **(Ans)**

(b) $I_r = \dfrac{I_{max}}{\sqrt{2}} = \dfrac{400 \text{ mA}}{\sqrt{2}} = 282.84$ mA

$V_r = 282.84 \times 0.1 = 28.28$ V **(Ans)**

(c) $\gamma = \sqrt{\dfrac{V_r^2 - V_D^2}{V_D^2}} = \sqrt{\dfrac{28.28^2 - 25.48^2}{25.48^2}} = 0.48$ **(Ans)**

(d) $\eta = \dfrac{254.81^2 \times 100}{282.84^2 \times 106} = 76.7\%$ **(Ans)**

(e) PIV = $15\sqrt{2} = 21.21$ V **(Ans)**

(f) Rating of the load resistance $= I_D^2 R_L = 254.81^2 \times 0.100 = 6.49$ W **(Ans)**

Usually, for safety, we choose R_L having a rating of $2 \times 6.49 = 13$ W **(Ans)**.

Chapter-2

One-Word Multiple-Choice Questions

M2.1 $\mu = -\partial V_{DS}/\partial V_{GS}$

M2.2 $g_m = \partial I_D/\partial V_{GS}$

M2.3 $r_d = \partial V_{DS}/\partial I_D$

M2.4 Inversion

M2.5 Quiescent

M2.6 Q-point

M2.7 Steady

M2.8 Load

M2.9 Fixed

M2.10 I_{CO}

M2.11 $\alpha I_E + I_{CO}$

M2.12 $I_C = g_m V_{BE}$

M2.13 $g_m = \dfrac{\partial I_C}{\partial V_{BE}}$

M2.14 $g_m = \left|\dfrac{I_C}{V_T}\right|$

M2.15 $S_I = \left|\dfrac{\partial I_C}{\partial I_{CO}}\right|$

M2.16 $S_V = \left|\dfrac{\partial I_C}{\partial V_{BE}}\right|$

M2.17 $S_V = \left|\dfrac{\partial I_C}{\partial \beta}\right|$

M2.18 $S_I = 1 + \dfrac{R_B}{R_E}.$

M2.19 $S_V = \dfrac{-1}{R_E}$

M2.20 Thermal runaway

M2.21 Current mirror

M2.22 Self biasing

M2.23 Voltage-divider biasing

M2.24 $r_b + r_\pi$

M2.25 $g_m r_\pi$

M2.26 $\dfrac{r_\pi}{r_\pi + r_\mu}$

M2.27 $\dfrac{1}{r_o} + \dfrac{1 + g_m r_\pi}{r_\pi + r_\mu}$

Unsolved Problems

P2.1 Using the relation $\alpha = I_C/I_E$,

we get $\alpha = I_C/I_E = 1/1.1 = 0.91$ **(Ans)**

P2.2 Using the relations $\beta = I_C/I_B$, and $\alpha = \beta/1 + \beta$,

we get $\beta = 1/0.01 = 100$, and $\alpha = 0.99$ **(Ans)**

P2.3 Using the relations $\beta = I_C/I_B$, and $\alpha = \beta/1 + \beta$, and $\gamma = 1 + \beta$,

we get $\beta = 1/0.01 = 100$, and $\alpha = 0.99$, and $\gamma = 101$ **(Ans)**

P2.4 Using the relation $\alpha = \Delta I_C/\Delta I_E$, we get $\alpha = I_C/I_E = 1/1.1 = 0.91$ **(Ans)**

P2.5 We get, by using the relation $\alpha = \Delta I_C/\Delta I_E$,

$$\alpha = \Delta I_C/\Delta I_E = (2 - 1)/(2.02 - 1.01) = 0.99 \ \textbf{(Ans)}$$

Configuration = *CBC* **(Ans)**

P2.6 The hybrid parameters can be computed using the basic relations

$$h_{ie} = \left[\frac{\Delta V_{BE}}{\Delta I_B}\right]_{V_{CE=\text{ constant}}} = \left[\frac{0.7 - 0.6}{(70 - 50) \times 10^{-6}}\right]_{V_{CE}=5\text{ V}} = \frac{0.1}{20 \times 10^{-6}} = 5\text{ k}\Omega \ \textbf{(Ans)}$$

$$h_{re} = \left[\frac{\Delta V_{BE}}{\Delta V_{CE}}\right]_{I_B=20} \left[\frac{0.7 - 0.6}{10 - 5}\right]_{I_B=55\,\mu\text{A}} = \frac{0.1}{5} = 20 \times 10^{-3} \ \textbf{(Ans)}$$

$$h_{fe} = \left[\frac{\Delta I_C}{\Delta I_B}\right]_{V_{CE=\text{ constant}}} = \left[\frac{(4 - 2) \times 10^{-3}}{(20 - 10) \times 10^{-6}}\right]_{V_{CE}=10\text{ V}} = \frac{2 \times 10^3}{10} = 200 \ \textbf{(Ans)}$$

$$h_{oe} = \left[\frac{\Delta I_C}{\Delta V_{CE}}\right]_{I_B=20} \left[\frac{(4 - 2)\text{ mA}}{10}\right]_{I_B=20\,\mu\text{A}} = \frac{2\text{ mA}}{10\text{ V}} = 0.2\text{ mS} \ \textbf{(Ans)}$$

P2.7 Use the relation $\beta = \Delta I_C / \Delta I_B$ to yield

$$\beta = \Delta I_C / \Delta I_B = (2 - 1)\,\text{mA} / (20 - 15)\,\mu\text{A} = 200 \textbf{ (Ans)}$$

Configuration = *CEC* **(Ans)**

P2.8 Use the relation $h_{rb} = \Delta V_{CB} / \Delta V_{EB}$ to yield

$$h_{rb} = \Delta V_{EB} / \Delta V_{CB} = (0.75 - 0.55) / (-5 + 10) = 40 \times 10^{-3} \textbf{ (Ans)}$$

Configuration = *CBC* **(Ans)**

P2.9 Use the relation $h_{re} = \Delta V_{CE} / \Delta V_{BE}$ to yield

$$h_{re} = \Delta V_{CE} / \Delta V_{BE} = (0.7 - 0.55) / (10 - 5) = 30 \times 10^{-3} \textbf{ (Ans)}$$

Configuration = *CEC* **(Ans)**

Chapter-3

One-Word Multiple-Choice Questions

M3.1 Abrupt

M3.2 Linearly graded

M3.3 Passivation

M3.4 Forward, Reverse

M3.5 Forward

M3.6 Cut-off, active and saturation

M3.7 Forward

M3.8 False

M3.9 Modulation

M3.10 Early

M3.11 Negative

M3.12 True

M3.13 False

M3.14 Low

M3.15 High

M3.16 Medium

M3.17 High

M3.18 Low

M3.19 Medium

M3.20 Low

M3.21 High

M3.22 High

M3.23 High

M3.24 Low

M3.25 High

M3.26 Low

M3.27 High

M3.28 $\beta = \dfrac{\alpha}{1 - \alpha}$

M3.29 $\gamma = \mathbf{1} + \beta$

M3.30 High

M3.31 Electric

M3.32 Cut-off

M3.33 Pinch-off

M3.34 Saturation

M3.35 Drain resistance

M3.36 Transconductance

M3.37 Amplification factor

M3.38 Quiescent

M3.39 *Q*-point

M3.40 Steady

M3.41 Load

M3.42 Fixed

M3.43 I_{CO}

M3.44 $I_C = g_m V_{BE}$

M3.45 $g_m = \dfrac{\partial I_C}{\partial V_{BE}}$

M3.46 V_T as $g_m = \left|\dfrac{I_C}{V_T}\right|$

M3.47 $S_I = \left|\dfrac{\partial I_C}{\partial I_{CO}}\right|$

M3.48 $S_V = \left|\dfrac{\partial I_C}{\partial V_{BE}}\right|$

M3.49 $S_V = \left|\dfrac{\partial I_C}{\partial \beta}\right|$

M3.50 $S_I = 1 + \dfrac{R_B}{R_E}$

M3.51 $S_V = \dfrac{-1}{R_E}$

M3.52 Thermal runaway

M3.53 Current mirror

M3.54 Self biasing

M3.55 Voltage-divider biasing

M3.56 $r_b + r_\pi$

M3.57 $g_m r_\pi$

M3.58 $\dfrac{r_\pi}{r_\pi + r_\mu}$

M3.59 $\dfrac{1}{r_o} + \dfrac{1 + g_m r_\pi}{r_\pi + r_\mu}$

M3.60 Stabilisation

M3.61 *Q*-point

M3.62 Collector.

M3.63 Blocking DC and passing AC

M3.64 $V_T = \dfrac{kT}{e}$

Unsolved Problems

P3.1 $S_I = \dfrac{(1+\beta)(R_B + R_E)}{R_B + (1+\beta) R_E} = \dfrac{(100+1)(100+1) \times 1 \times 10^3}{100 \times 10^3 + (100+1) \times 1 \times 10^3}$

$$= \frac{101 \times 101 \times 10^3}{201 \times 10^3} = 50.75 \textbf{ (Ans)}$$

P3.2 $S_I = \dfrac{(1+\beta)(R_B + R_E)}{R_B + (1+\beta) R_E} = \dfrac{(100+1)(10+1) \times 1 \times 10^3}{10 \times 10^3 + (100+1) \times 1 \times 10^3}$

$$= \frac{101 \times 11 \times 10^3}{110 \times 10^3} = 10.1 \textbf{ (Ans)}$$

P3.3 $S_V = \dfrac{-\beta}{R_B + (1+\beta) R_E}$

$$= \frac{-100}{10 \times 10^3 + (100+1) \times 1 \times 10^3} = 0.9 \times 10^{-3} \textbf{ (Ans)}$$

P3.4 $$S_\beta = \frac{(R_B + R_E) I_{C1}}{\beta_1 [R_B + (1+\beta_2) R_E]} = \frac{(10+1) \times 10^3 \times 1 \times 10^{-3}}{100\left[10 \times 10^3 + (105+1) \times 1 \times 10^3\right]}$$

$$= \frac{11}{100 \times 116 \times 10^{-3}} = 0.948 \textbf{ (Ans)}$$

Chapter-4

One-Word Multiple-Choice Questions

M4.1 Electric

M4.2 Cut-off

M4.3 Pinch-off

M4.4 Saturation

M4.5 Drain resistance

M4.6 Transcoductance

M4.7 Amplification factor

M4.8 $\mu = -\partial V_{DS} / \partial V_{GS}$

M4.9 $g_m = \partial I_D / \partial V_{GS}$

M4.10 $r_d = \partial V_{DS} / \partial I_D$

M4.11 Inversion

M4.12 Shockley

M4.13 $I_A = \dfrac{I_{CO}}{1 - (\alpha_p + \alpha_n)}$

M4.14 ∞

M4.15 Blocking

M4.16 On state

M4.17 Latching current

M4.18 Holding

M4.19 Negative

M4.20 Switching

M4.21 Intrinsic stand-off

M4.22 Relaxation oscillator

M4.23 Conductivity modulation

M4.24 Peak

M4.25 Valley

Unsolved Problems

P4.1 We have from definitions, the drain resistance given by,

$$r_d = \left[\frac{\Delta V_{DS}}{\Delta I_D}\right]_{V_{GS}=\text{const.}}$$

Substituting values from the table, we find

$$r_d = \left[\frac{6-4}{(2-1) \times 10^{-3}}\right]_{V_{GS}=2\text{ V}} = 2 \text{ k}\Omega \textbf{ (Ans)}$$

The transconductance is given by

$$g_{\text{m}} = \left[\frac{\Delta I_{\text{D}}}{\Delta V_{\text{GS}}}\right]_{V_{\text{DS}}=\text{const.}}$$

Substituting values, we get

$$g_{\text{m}} = \left[\frac{(4-3)\ \text{mA}}{-2-(-3)}\right]_{V_{\text{DS}}=10\ \text{V}} = 1.0\ \text{mS}\ \textbf{(Ans)}$$

The amplification factor is given by

$$\mu = -\left[\frac{\Delta V_{\text{DS}}}{\Delta V_{\text{GS}}}\right]_{V_{\text{DS}}=\text{const.}}$$

Substituting values, we get

$$\mu = -\left[\frac{10-8}{-3-(-2)}\right]_{I_{\text{D}}=3\ \text{mA}} = \frac{2}{1} = 2\ \textbf{(Ans)}$$

P4.2 $\mu = -\left[\frac{\Delta V_{\text{DS}}}{\Delta V_{\text{GS}}}\right]_{V_{\text{DS}}=\text{const.}} = -\left[\frac{15-5}{-1-0}\right] = \frac{10}{1} = 10\ \textbf{(Ans)}$

P4.3 $g_{\text{m}} = \left[\frac{\Delta I_{\text{D}}}{\Delta V_{\text{GS}}}\right]_{V_{\text{DS}}=\text{const.}} = -\left[\frac{4-2}{-2-(-3)}\right] = \frac{2\ \text{mA}}{1} = 2\ \text{mS}\ \textbf{(Ans)}$

P4.4 $r_{\text{d}} = \left[\frac{\Delta V_{\text{DS}}}{\Delta I_{\text{D}}}\right]_{V_{\text{DS}}=\text{const.}} = -\left[\frac{15-10}{4-2}\right] = \frac{5}{2} = 2.5\ \text{k}\Omega\ \textbf{(Ans)}$

P4.5 The frequency of oscillation is given by

$$T = RC\ln\frac{1}{1-\eta} = 6\times10^{3}\times0.01\times10^{-6}\ln\frac{1}{1-0.7} = 72.24\ \mu s$$

Hence frequency of oscillation is given by

$$f_{\text{o}} = \frac{1}{T} = \frac{1}{72.24\ \mu s} = 13.84\ \text{kHz}\ \textbf{(Ans)}$$

P4.6 We have the intrinsic stand-off ratio of the UJT given by

$$\eta = \frac{R_{\text{B1}}}{R_{\text{B1}}+R_{\text{B2}}} = \frac{6\times10^{3}}{(6+4)\times10^{3}} = 0.6$$

The frequency of oscillation is given by

$$T = 6 \times 10^3 \times 0.001 \times 10^{-6} \ln \frac{1}{04} = 5.5\ \mu s$$

Hence frequency of oscillation is given by

$$f_o = \frac{1}{T} = \frac{1}{5.5\ \mu s} = 181.9 \text{ kHz } \textbf{(Ans)}$$

P4.7 We have

$$I_A = \frac{I_{CO}}{1-(\alpha_p + \alpha_n)} = \frac{1.5 \times 10^{-10}}{1-(2 \times 0.498)} = 37.5 \text{ nA } \textbf{(Ans)}$$

P4.8 We have

$$I_A = \frac{I_{CO}}{1-(\alpha_p + \alpha_n)} = \frac{1.5 \times 10^{-10}}{1-(2 \times 0.4995)} = 6.67 \text{ A } \textbf{(Ans)}$$

[**Note:** It can be seen that, when the sum of α_p and α_n approaches unity, the value of anode current approaches infinity].

Chapter-5

One-Word Multiple-Choice Questions

M5.1 Stabilisation

M5.2 *Q*-point

M5.3 Collector

M5.4 Passing AC

M5.5 $V_T = \frac{kT}{e}$

M5.6 Gain

M5.7 Transition frequency

M5.8 f_T

M5.9 Transition

M5.10 Lower or Upper

M5.11 Bandwidth

M5.12 3-dB bandwidth

M5.13 Unity

M5.14 Unity

M5.15 Unity

M5.16 Barkhausen

M5.17 $AB = 1\angle 0°$

M5.18 $AB = 1\angle 360°$.

M5.19 1. Unity
2. 0 or 360°

M5.20 $f = \frac{1}{2\pi RC}$

M5.21 $f = \frac{1}{2\pi RC\sqrt{6}}$

M5.22 $f = \frac{\sqrt{6}}{2\pi RC}$

M5.23 −29

M5.24 −29

M5.25 3

M5.26 *RLC*

Unsolved Problems

P5.1 We have

$$\frac{I_{D1}}{I_{D2}} = \frac{W_1/L_1}{W_2/L_2}$$

Substituting values, we get

$$\frac{2}{1} = \frac{W_1 L_2}{W_2 L_1} = \frac{1 \times L_2}{2 \times L_1}$$

Then the ratio of lengths of the channel is given by

$$\frac{L_2}{L_1} = 2 \textbf{ (Ans)}$$

P5.2 Substituting values, we get the ratio of drain currents

$$\frac{I_{D1}}{I_{D2}} = \frac{W_1 L_2}{W_2 L_1} = \frac{1 \times 1}{2 \times 2} = \frac{1}{4} \textbf{ (Ans)}$$

P5.3 The expression for the transconductance of a BJT is

$$g_m = \left| \frac{I_C}{V_T} \right|$$

If we assume that thermal voltage $V_T = 0.025$ V at room temperature, then

$$g_m = \frac{1 \text{ mA}}{0.025} = 40 \text{ mS } \textbf{(Ans)}$$

P5.4 Substituting values in the expression $\beta = g_m r_\pi$, we obtain

$$\beta = g_m r_\pi = 40 \times 10^{-3} \times 3 \times 10^3 = 120 \textbf{ (Ans)}$$

P5.5 Substituting values in the expression $\beta = g_m r_\pi$, we obtain

$$\beta = g_m r_\pi = 40 \times 10^{-3} \times 1 \times 10^3 = 40 \textbf{ (Ans)}$$

P5.6 Substituting values in the expression, $h_{re} = \dfrac{r_\pi}{r_\mu + r_\pi}$ we obtain

$$h_{re} = \frac{1}{3 + 1000} = 9.97 \times 10^{-4} \textbf{ (Ans)}$$

P5.7 Substituting values in the expression, $Y_o = \dfrac{1}{r_o} + \dfrac{1 + g_m r_\pi}{r_\mu + r_\pi}$ we obtain

$$Y_o = \frac{1}{1000 \times 10^3} + \frac{1 + 200}{1003 \times 10^3} = 0.2 \text{ mS } \textbf{(Ans)}$$

$$f_T = h_{fe} f_\beta = 100 \times 395,000 = 39.5 \text{ MHz } \textbf{(Ans)}$$

$$A_v = \frac{\mu R_L}{r_d + R_L} = \frac{(g_m r_d) R_L}{r_d + R_L} = \frac{2 \times 10 \times 5}{10 + 5} = 6.67 \textbf{ (Ans)}$$

P5.8 The frequency of oscillation is given by

$$f_o = \frac{1}{2\pi RC} = \frac{1}{2\pi \times 1 \times 10^3 \times 0.01 \times 10^{-6}} = 15.92 \text{ kHz } \textbf{(Ans)}$$

P5.9 The frequency of oscillation is given by

$$f_o = \frac{1}{2\pi RC} = \frac{1}{2\pi \times 3.3 \times 10^3 \times 0.005 \times 10^{-6}} = 9.65 \text{ kHz } \textbf{(Ans)}$$

P5.10 The frequency of oscillation is given by

$$f_o = \frac{1}{2\pi RC\sqrt{6}} = \frac{1}{2\pi \times 2.7 \times 10^3 \times 0.001 \times 10^{-6}\sqrt{6}} = 24.065 \text{ kHz } \textbf{(Ans)}$$

P5.11 The frequency of oscillation is given by

$$f_o = \frac{1}{2\pi\sqrt{LC}}$$

where $L = L_1 + L_2 = 1\ \mu\text{H} + 3\ \mu\text{H} = 4\ \mu\text{H}$. Substituting values obtained, we get

$$f_o = \frac{1}{2\pi\sqrt{LC}} = \frac{1}{2\pi \times \sqrt{4 \times 10^{-6} \times 100 \times 10^{-12}}} = 7.96 \text{ MHz } \textbf{(Ans)}$$

Gain of the amplifier to be used is given by

$$f_o = \frac{L_2}{L_1} = \frac{3\ \mu\text{H}}{1\ \mu\text{H}} = 3 \textbf{ (Ans)}$$

P5.12 The frequency of oscillation is given by

$$f_o = \frac{1}{2\pi\sqrt{LC}}$$

where $L = L_1 + L_2 \pm 2M = 1\ \mu\text{H} + 3\ \mu\text{H} + 10\ \mu\text{H} = 14\ \mu\text{H}$. Substituting yields (assuming positive value for M; for negative value of M, f_o will be different).

$$f_o = \frac{1}{2\pi\sqrt{LC}} = \frac{1}{2\pi \times \sqrt{14 \times 10^{-6} \times 100 \times 10^{-12}}} = 4.25 \text{ MHz } \textbf{(Ans)}$$

P5.13 The frequency of oscillation is given by

$$f_o = \frac{1}{2\pi\sqrt{LC}}$$

where $L = L_1 + L_2 \pm 2M = 3\ \mu\text{H} + 5\ \mu\text{H} - 6\ \mu\text{H} = 2\ \mu\text{H}$. Substituting yields

$$f_o = \frac{1}{2\pi\sqrt{LC}} = \frac{1}{2\pi \times \sqrt{2 \times 10^{-6} \times 100 \times 10^{-12}}} = 11.25\ \text{MHz}\ \textbf{(Ans)}$$

P5.14 The frequency of oscillation is given by

$$f_o = \frac{1}{2\pi\sqrt{LC}}$$

where $C = \dfrac{C_1 C_2}{C_1 + C_2}$. Substituting $C_1 = 100$ pF, $C_2 = 100$ pF, we get

$$C = \frac{100 \times 100 \times 10^{-24}}{100 \times 10^{-12} + 100 \times 10^{-12}} = \frac{100 \times 100 \times 10^{-24}}{200 \times 10^{-12}} = 50 \times 10^{-12}\ \text{F}$$

Then

$$f_o = \frac{1}{2\pi\sqrt{LC}} = \frac{1}{2\pi \times \sqrt{5 \times 10^{-6} \times 50 \times 10^{-12}}} = 10.07\ \text{MHz}\ \textbf{(Ans)}$$

Gain of the amplifier to be used is

$$A = \frac{C_1}{C_2} = 1\ \textbf{(Ans)}$$

P5.15 The frequency of oscillation is given by

$$f_o = \frac{1}{2\pi\sqrt{LC}}$$

where $C = \dfrac{C_1 C_2}{C_1 + C_2}$. Substituting $C_1 = 400$ pF, $C_2 = 100$ pF, we get

$$C = \frac{400 \times 100 \times 10^{-24}}{400 \times 10^{-12} + 100 \times 10^{-12}} = \frac{400 \times 100 \times 10^{-24}}{500 \times 10^{-12}} = 80 \times 10^{-12}\ \text{F}$$

Then

$$f_o = \frac{1}{2\pi\sqrt{LC}} = \frac{1}{2\pi \times \sqrt{5 \times 10^{-6} \times 80 \times 10^{-12}}} = 7.95\ \text{MHz}\ \textbf{(Ans)}$$

Gain of the amplifier to be used is

$$A = \frac{C_1}{C_2} = 4\ \textbf{(Ans)}$$

Chapter-6

One-Word Multiple-Choice Questions

M6.1 Virtual short

M6.2 180°

M6.3 0°

M6.4 $1 + (R_2/R_1)$

M6.5 $-R_2/R_1$

M6.6 Voltage-follower

M6.7 High

Unsolved Problems

P6.1 The output voltage of a opamp summer is given by

$$-V_o = \frac{R_F}{R_1}V_1 + \frac{R_F}{R_2}V_2 + \frac{R_F}{R_3}V_3 + \frac{R_F}{R_4}V_4$$

Substituting given values yields

$$-V_o = \frac{8}{8} \times 1 + \frac{8}{4} \times 1 + \frac{8}{2} \times 1 + \frac{8}{1} \times 1 = 15 \text{ V } \textbf{(Ans)}$$

P6.2 The out put voltage of an opamp inverter is given by

$$V_o = -V_i = -10 \text{ V } \textbf{(Ans)}$$

P6.3 The out put voltage of a opamp inverter is given by

$$V_o = -\frac{1}{RC}\int V_i dt$$

Since the period of the applied wave is given by

$$T = 1/f$$

where frequency f is obtained from the expression $\sin(200\pi t) = \sin(2\pi f t)$ as 100 Hz, we obtain

$$T = \frac{1}{f} = \frac{1}{100} = 0.01 \text{ s}$$

For an integrator, using the design rule that $RC \gg T$, we choose RC such that

$$RC = 100T = 1 \text{ s}$$

Assuming

$$C = 1 \text{ F } \textbf{(Ans)}$$

we select

$$R = 1\ \text{M}\Omega\ \textbf{(Ans)}$$

The output voltage therefore becomes

$$V_o = \int 10\ \sin(200\pi t)\ dt = \frac{10\ \cos(200\pi t)}{200\pi} = 0.016\ \cos(200\pi t)\ \textbf{(Ans)}$$

P6.4 The out put voltage of an opamp differential amplifier is given by

$$V_o = \frac{R_2}{R_1}(V_2 - V_1) = \frac{100}{10}(4 - 2) = 10 \times 2 = 30\ \text{V}\ \textbf{(Ans)}$$

Chapter-7

One-Word Multiple-Choice Questions

M7.1 $h = \frac{\lambda}{2}$

M7.2 $v(t) = V_c \cos(\omega_c t) + \frac{mV_c}{2}\cos(\omega_c + \omega_m)t + \frac{mV_c}{2}\cos(\omega_c - \omega_m)t$

M7.3 $v(t) = V_c \sin(\omega_c t + m_f \sin \omega_m t)$

M7.4 $v(t) = V_c \sin\left(\omega_c t + m_p \sin \omega_m t\right)$

M7.5 $m_a = \frac{V_{max} - V_{min}}{V_{max} + V_{min}}$ or $k\frac{V_m}{V_c}$

M7.6 $m_f = \frac{k_f V_m}{\omega_m} = \frac{\Delta f}{f_m}$

M7.7 $m_f = k_p V_m$

M7.8 Crystal

M7.9 Intermediate (or IF)

M7.10 Two

M7.11 Reactance

M7.12 Armstrong

M7.13 FM

M7.14 Three

M7.15 ±75 kHZ

M7.16 Binary digit

M7.17 Binary number system

M7.18 Octal number system

M7.19 Hexadecimal number system

M7.20 $(10)_{10}$

M7.21 $(1010)_2$

M7.22 $(111111111111)_2$

M7.23 $(111111111111)_2 = (7777)_8$

M7.24 $(977)_{BCD}$

M7.25 $(001100100)_2 = (144)_8$

Unsolved Problems

P7.1 Height of a resonant antenna is given by

$$h = \frac{\lambda}{2}$$

where $\lambda = \frac{c}{f} = \frac{3 \times 10^8}{10^6} = 3 \times 10^2$

Hence antenna height is given by

$$h = \frac{300}{2} = 150 \text{ m } \textbf{(Ans)}$$

P7.2

$$\lambda = \frac{c}{f} = \frac{3 \times 10^8}{10^9} = 30 \text{ cm}$$

Hence antenna height is given by

$$h = 30/2 = 15 \text{ cm } \textbf{(Ans)}$$

P7.3 The amplitude-modulation index is given by

$$m_a = \frac{V_m}{V_c} = \frac{50}{100} = 0.7$$

The expression for the modulated signal is given by

$$\begin{aligned} v_t &= V_c \cos \omega_c t + \frac{mV_c}{2} \cos(\omega_c + \omega_m)\, t + \frac{mV_c}{2} \cos(\omega_c - \omega_m)\, t \\ &= 100 \cos\left(2\pi \times 10^6\right) t + 35 \cos\left(2\pi \times 1.01 \times 10^6\right) t + 35 \cos\left(2\pi \times 0.99 \times 10^6\right) t \\ &= 100 \cos\left(2\pi \times 10^6\right) t + 35 \cos\left(2.02\pi \times 10^6\right) t + 35 \cos\left(1.98\pi \times 10^6\right) t \ \textbf{(Ans)} \end{aligned}$$

P7.4

$$\omega_t = \omega_c + (\Delta f / f_m)\ \sin \omega_c t$$

Substituting values, we get

$$\begin{aligned} \omega_t &= 2\pi \times 10^6 + \left(75 \times 10^3/10^4\right) \sin\left(2\pi \times 10^6\right) t \\ &= 2\pi \times 10^6 + 7.5 \sin\left(2\pi \times 10^6\right) t \end{aligned}$$

The expression of the modulated wave then is

$$v_t = 100 \sin\left[2\pi \times 10^6 + 7.5 \sin\left(2\pi \times 10^6\right) t\right] \ \textbf{(Ans)}$$

P7.5 Modulation index is given by

$$m_f = \frac{k_{f0} V_m}{\omega_m} \frac{1000 \times 70}{2\pi \times 10^4} = 1.11$$

The expression of the modulated wave then is

$$v_t = 100 \sin\left[2\pi \times 10^6 + 1.11 \sin\left(2\pi \times 10^6\right) t\right] \ \textbf{(Ans)}$$

P7.6 By intuition, we may write the binary equivalent of 100 as

$$(100)_{10} = (64)_{10}+(32)_{10}+(4)_{10} = 2^6+2^5+2^2 = 100,000+100,00+100 = 110,100 \textbf{ (Ans)}$$

P7.7 $(0.101111)_2 = 1\times 2^{-1} + 0\times 2^{-2} + 1\times 2^{-3} 1\times 2^{-4} 1\times 2^{-5}$

$$= \frac{1}{2} + \frac{1}{8} + \frac{1}{16} + \frac{1}{32} = \frac{23}{32} \textbf{ (Ans)}$$

P7.8

Number	0.2315	0.4630	0.9260	0.8520	0.7040
×2	0.4630	0.9260	0.8520	0.7040	0.408
Remainder	0	0	1	1	1

Number	0.408	0.816	0.632	0.264	0.528
×2	0.816	0.632	0.264	0.528	0.056
Remainder	0	1	1	0	1

Tabulation is made as shown in the tables. It can be seen that Table 2 is a continuation of Table 1. The entries in the tables continue till the last column of the last table shows "0" as its entry. The converted binary, then, is

$$(0.2315)_{10} = (0.0011101101)_2 \textbf{ (Ans)}$$

P7.9

Decimal number	6	51	414	3312
Number ÷ 8		6	51	414
Balance (octal) equivalent	6	3	6	0

$$(3312)_{10} \equiv (6360)_8$$

P7.10

Octal number	3	3	1	2
Binary equivalent	011	011	001	010

From the table, we get

$$(3312)_8 = (11011001010)_2$$

P7.11

Binary number	110	111	100	011
Octal equivalent	6	7	4	3

$$(110111100100011)_2 = (6743)_8$$

P7.12

Binary number	1101	1110	0100	0110
Hex equivalent	D	E	4	6

From table, we obtain

$$(1101111001000110)_2 \equiv (DE46)_{16}$$

P7.13 From table, we obtain

Hex number	C	C	A	B
Binary equivalent	1100	1100	1010	1011

$$(\text{CCAB})_{16} \equiv (1100110010101011)_2$$

P7.14

Decimal number	2	6	9
BCD code	0010	0110	1001

$$(269)_{10} = (001001101001)_{\text{BCD}}$$

P7.15

Decimal number	3	2	1	9	8
BCD code	0011	0110	0001	1001	1000

$$(32198)_{10} = (00110110000110011000)_{\text{BCD}}$$

P7.16

BCD code	1001	1000	0101	0001	0011
Decimal number	9	8	5	1	3

The decimal number is 98,513

Chapter-8

One-Word Multiple-Choice Questions

M8.1 Boolean

M8.2 AND, OR and NOT

M8.3 AND

M8.4 OR

M8.5 CMOS NOR

M8.6 CMOS NAND

M8.7 Negligible

M8.8 Open

M8.9 Universal

M8.10 NAND and NOR

M8.11 Transistor-transistor logic

M8.12 Emitter-coupled logic

M8.13 Resistor-transistor logic

M8.14 Diode-transistor logic

M8.15 Integrated-injection logic

M8.16 Idempotency

M8.17 Idempotency

M8.18 Absorption

M8.19 Absorption

M8.20 Consensus

Unsolved Problems

P8.1 1. Commutative Law

$$a + b = b + a$$
$$ab = ba$$

2. Associative Law

$$a + (b + c) = (a + b) + c$$
$$a(bc) = (ab)c$$

3. Distributive Law

$$w(x + y + z) = wx + wy + wz$$

4. Absorption Laws

$$a + abc = a$$
$$a(a + b + c) = a$$

5. Absorption Laws

$$a + a'b = a + b$$

$$a\,(a' + b) = ab$$

6. Absorption Law

$$a' + ab = a' + b$$

$$a'\,(a + b) = a'b$$

7. Idempotency

$$a + a = a$$

$$aa = a$$

P8.2 The sum term of a full adder is given by

$$\begin{aligned} S &= A\overline{BC} + \overline{A}B\overline{C} + \overline{AB}C + ABC \\ &= A\left(\overline{BC} + BC\right) + \overline{A}\left(B\overline{C} + \overline{B}C\right) = A\overline{X} + \overline{A}X \end{aligned}$$

where we define $X = B\overline{C} + \overline{B}C$ and $\overline{X} = BC + \overline{BC}$. It can be seen that $X = B\overline{C} + \overline{B}C$ represents a EXOR (half-adder) and $\overline{X} = BC + \overline{BC}$ represents an EXNOR function. Thus we find that the full-adder is equivalent to the sum of an EXOR function and an EXNOR function. In effect we state that: **One full adder = Sum of two half-adders**

P8.3 First write the given function in the form. The double barring does not change the value of the given function. Thus

$$\overline{\overline{Y}} = \overline{\overline{A + B + C}}$$

Now, retain the topmost bar and apply DeMorgan's law to the bottom bar. This yields the expression

$$Y = \overline{\overline{A}\,\overline{BC}}$$

This expression can be implemented using the circuit shown below.

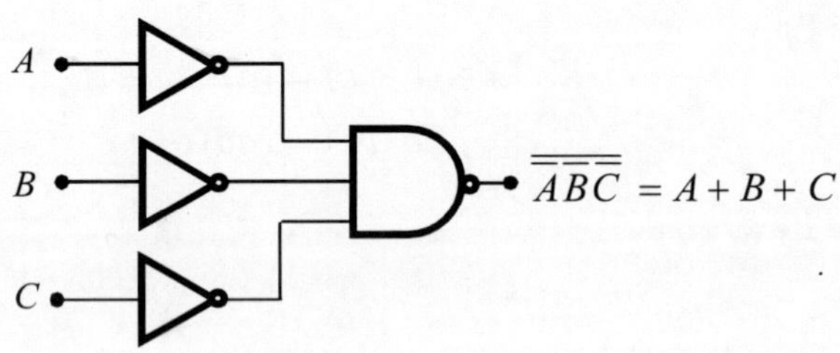

x	y	z	$(x + y)$	$(x + y) + z$	$(y + z)$	$x + (y + z)$
0	0	0	0	0	0	0
0	0	1	0	1	1	1
0	1	0	1	1	1	1
0	1	1	1	1	1	1
1	0	0	1	1	1	1
1	0	1	1	1	1	1
1	1	0	1	1	1	1
1	1	1	1	1	1	1

Comparing the columns designated $(x + y) + z$ and $x + (y + z)$, we find that the entries are the same.

INDEX